T0039084

GSN—The Goal Structuring Notation

John Spriggs

GSN—The Goal Structuring Notation

A Structured Approach to Presenting Arguments

Springer

John Spriggs
Hayling Island
UK
e-mail: GSN.Book@Gmail.com

ISBN 978-1-4471-2311-8 e-ISBN 978-1-4471-2312-5
DOI 10.1007/978-1-4471-2312-5
Springer London Dordrecht Heidelberg New York

British Library Cataloguing in Publication Data
A catalogue record for this book is available from the British Library

Library of Congress Control Number: 2011940782

Cover design: eStudio Calamar S.L.

Printed on acid-free paper

Springer is part of Springer Science+Business Media (www.springer.com)

To my wife Bonnie, who once said that every book should be dedicated to someone special

Preface

Motivation

I have often been asked over the last decade, or so, 'Can you recommend a book on GSN?' My answer has always (well, up to now at least) been, 'No'.

This was not because I found all such books unworthy; rather it is because I had not found any books on the subject. When someone asked me the question just after I had given a tutorial on GSN, I gave my stock answer. They came straight back and said that I should write one.

So, after much delay, here it is.

This book is both for the aspiring writer of arguments in GSN and for those who wish to read, review and understand them. It presents a clear exposition of the notation; it is a narrative, starting from first principles and showing why each symbol is needed and how it should be used. Each chapter builds on what has gone before. This is not a learned dissertation; it is a practitioners' guide to GSN, the Goal Structuring Notation.

Important Notes

The purpose of this book is to give an introduction to the Goal Structuring Notation and its use in representing arguments. It does not present any ready-to-use arguments for redeployment elsewhere. In particular, it does not present any safety arguments for re-use. A safety argument depends not only on the detail of the thing argued about, but also upon the environment in which it is used. A safety argument requires careful thought and analysis of the pertinent risks. If you ever re-use parts of such arguments, or claim benefit from someone else's argument, add an explicit justification of why you think it is appropriate for use in the new context.

The graphic illustrations appearing in this book are conceptual and were developed strictly to support discussions in the text. They are not intended to be used as the basis of assurance arguments. A number of the examples refer to legislation or to regulatory requirements, but note that this book does not offer legal advice; neither does it recommend means of compliance.

The author of this book, his past and present employers and clients, and his publishers share neither liability nor responsibility for any loss or damaged caused, or alleged to have been caused, by information obtained from this book.

Acknowledgments

The Goal Structuring Notation was formalised for use as a means of presenting safety arguments by Tim Kelly at York University.

Dr. Kelly is still supervising GSN development by his students and colleagues, both at York University and in industry. He has a web-page listing his many publications on the topic at http://www-users.cs.york.ac.uk/~tpk/.

I largely based the extended example of Chaps. 10, 11 and 12 of this book on a presentation that I gave on 20th January 2009 to the Solent Branch of the Safety and Reliability Society, SaRS, http://www.sars.org.uk/. Thank you to them for giving me the opportunity to judge the interest in this topic and to see if the example was straightforward enough to be illuminating without being so simple as to be dull. Note: I am indebted to The South Downs Planetarium and Science Centre in Chichester, UK, for introducing me to the subject of this example. They have a web-site at http://www.southdowns.org.uk/sdpt/.

Of course, many thanks are also due to Anthony Doyle, Christine Velarde, Claire Protherough, Grace Quinn and all their colleagues at Springer who made this book possible.

July 2011 John Spriggs

Contents

Chapter 1
Introduction

Abstract This introductory chapter presents the purpose of the book and the purpose of GSN, the notation described herein. It identifies the intended audience; how they could use the book to greatest advantage; and summarizes the content.

1.1 What is the Purpose of This Book?

This book describes the Goal Structuring Notation (GSN), which you can use to represent arguments. This is not argument as in having a fight over something; it is a process of reasoning, you are 'making a case'.

This book is both for the aspiring writer of arguments in GSN and for those who wish to read, review and understand them. It presents a clear exposition of the notation, starting from first principles and showing why each symbol is needed and how it should be used.

GSN was originally proposed as a means of representing system requirements decomposition, apportionment and means of achievement. The Goal is the desired outcome. The notation, and processes facilitating its use in representing arguments, were further developed (initially in the EPSRC-funded ASAM-II project) for systems safety assurance (Kelly 1998), but its application is in no way restricted to such use. We do not have to confine ourselves to arguments starting with "The System is Safe". We can argue, for example: "The defendant is guilty", or "It will rain here next week", or even, "I deserve a pay rise".

You may have picked up this book with a view to being able to read and critically review arguments expressed in GSN; presumably, after that last example, you now also want to learn how to create compelling arguments using GSN...

J. Spriggs, *GSN—The Goal Structuring Notation*,
DOI: 10.1007/978-1-4471-2312-5_1,

1.2 What is the Purpose of GSN; Who Would Use It?

GSN provides us with a set of symbols and rules for their use. These symbols enable us to draw diagrams of arguments.

> An argument is a connected series of statements intended to establish a proposition
> (Python 1989)

GSN makes these connections explicit; a diagram of an argument using GSN will thus be clearer than the text equivalent.

You use an argument to persuade others of the validity of some claim. GSN allows you to present your argument more clearly, as a series of diagrams, than you could in an unstructured text. Presenting your argument clearly is more likely to persuade.

The notation gives you more than just a means to persuade, however; it allows you to check your arguments. You can readily identify any weak, i.e. less than compelling, parts of the argument so that they can be "beefed-up". GSN is therefore a useful notation to use when developing your arguments even if it would be inappropriate to present the final version as a diagram, for example when asking for the pay rise mentioned previously.

Anybody wishing to support or refute a claim can use GSN. It is of particular use when an argument has to be formally presented in order to persuade others, for example an industry Regulator, when claiming, "It meets regulatory requirements", or a Planning Enquiry when claiming that, "The proposed development is inappropriate".

GSN can also be used to highlight what is missing from a scheme or undertaking; I give an example (in Chap. 5) of a Business Plan. You can develop an argument claiming that your plan is complete, from which it is easy to see what you still have to do to finish it properly. The same process can be followed if you have made an undertaking to someone; you develop an argument that you have done everything you promised and find what you've missed, if anything, on your side of the bargain.

Such arguments of completeness can also be used as the basis of checklists that may be used by a quality system auditor or a GSN reviewer, for example.

1.3 How Should I Use This Book?

Well; read it…

I wrote the book to be read in sequence from front to back. I have presented the information incrementally. It is a narrative, with each chapter building on what has gone before. Clear guidance and simple examples are given of each piece of the notation as it is introduced.

I suggest you read this book all through once and then use it as a reference to dip into as you construct your own arguments. That said, if you already know some of the notation, it should be easy to go straight into a chapter to find the detail that you want. To facilitate this, each chapter begins with an abstract in which I summarise the new concepts introduced therein. The last chapter is a summary of the notation for revision or quick reference.

This introductory chapter is the only one without diagrams. Each subsequent chapter presents illustrated examples to demonstrate the use of the symbols that it introduces, usually in combination with those from earlier chapters. These examples are, of necessity, very simple in early chapters, but some are revisited in later chapters to use the newly-introduced symbols to get a 'better' result.

Once sufficient notation is to hand, I give an example that has a chapter of its own and serves as a vehicle to introduce further new concepts and guidance over subsequent chapters. Note that I intentionally chose the subject matter of this example to be unfamiliar to the majority of readers. This is so that you can concentrate on the use of the notation, rather than worrying about my over-simplifications.

I have provided a subject index at the back of the book that shows on which page you can find a description of each symbol, and where any phrasing and labelling conventions are set out. It also lists the examples and main guidance topics so that you will be able to go back to the one you need.

Throughout the text, I have used leading capitals when referring to part of the notation, e.g. a Goal, in order to distinguish it from the word used in a conventional sense, e.g. a goal as a target to attain.

Note that this book is not a learned dissertation. I am not presenting research, or a theory of argumentation, rather I am providing a practitioners' guide to the GSN. Formal references are thus only provided where required to give credit, or to acknowledge the source of quotations.

1.4 Important Note

As previously stated, the purpose of this book is to give an introduction to the Goal Structuring Notation and its use in representing arguments. It does not present ready-to-use arguments for redeployment elsewhere. In particular, it does not present any safety arguments for re-use. A safety argument depends not only on the detail of the thing argued about, but also upon the environment in which it is used. A safety argument requires careful thought and analysis of the pertinent risks. If you ever re-use parts of such arguments, or claim benefit from someone else's argument, add an explicit justification of why you think it is appropriate for use in the new context.

The graphic illustrations appearing in this book are conceptual and were developed strictly to support discussions in the text. They are not intended to be used as the basis of assurance arguments. A number of the examples refer to legislation or to regulatory requirements, but note that this book does not offer legal advice; neither does it recommend means of compliance.

The author of this book, his past and present employers and clients, and his publishers share neither liability nor responsibility for any loss or damaged caused, or alleged to have been caused, by information obtained from this book.

1.5 Other Notations

There are, as they may say on the BBC, other notations available for presenting argumentation and reasoning. In my opinion, none of these are as expressive as Goal Structuring Notation and they can need a lot of text to support the diagrams.

If you want to explore these alternatives, I suggest you start with Wigmore Charts (Wigmore 1913), Toulmin Diagrams (Toulmin 1958), or Claims Argument Evidence Trees (Adelard 2002).

1.6 Questions and Problems

All chapters, other than this and the last, contain both Questions and Problems so that you can reflect on and try out what you have learned. The distinction is that (most) Questions consider the next logical step in the development, and so tend to be answered by a succeeding chapter, whereas Problems look at applying what has been covered so far. Answers to the Problems can be found at the back of the book. It is worth observing that many of the Problems have no single correct answer, my answer is a suggested approach, occasionally with alternatives.

There is perhaps one question that should be answered here: "Why is the main title of this book a set of initials?" This may be particularly puzzling to those who know me, and are aware that I rarely use acronyms or initialisms. The answer leads us on to another question that will be answered in the next Chapter.

It is usual in the literature, and in documented assurance arguments, to see this notation referred to just as "GSN" so, there currently being no other books on the subject, I used that as my title. Unfortunately, a significant proportion of writers go on to say that GSN is the "Goal Structured Notation". This is wrong. GSN is the Goal Structuring Notation. It has that name because it is a notation for structuring Goals.

So, what is a Goal? That question is answered in Chap. 2.

References

Emmet L, Cleland G (2002) Graphical notations, narratives and persuasion: a pliant systems approach to hypertext tool design, in hypertext 2002. In: Proceedings of the 13th ACM conference on hypertext and hypermedia, Association for Computing Machinery

Kelly TP (1998) Arguing safety—a systematic approach to managing safety cases (doctoral dissertation). Department of Computer Science, University of York

Chapman G, Cleese J, Gilliam T, Idle E, Jones T, Palin M (1989) The argument clinic, in volume 2 of Monty Python's flying circus ~ Just the Words, Methuen

Toulmin S (1958) The uses of argument. Cambridge University Press, Cambridge

Wigmore J (1913) The principles of judicial proof. Little Brown & Co, Boston

Chapter 2
Goals

Abstract The Goal Structuring Notation is a notation for structuring Goals but, what is a Goal and why would we want to structure them?

When using the notation to represent arguments, the Goals represent claims. We can use Goal Structures to persuade others of the truth of our claims. This chapter introduces the symbol and text convention to be used to represent Goals in these structures.

2.1 What is a Goal?

Look up "goal" in a dictionary and you are likely to find something along the lines of: A rectangular opening through which, in order to score, a soccer player must pass the ball, see Fig. 2.1.

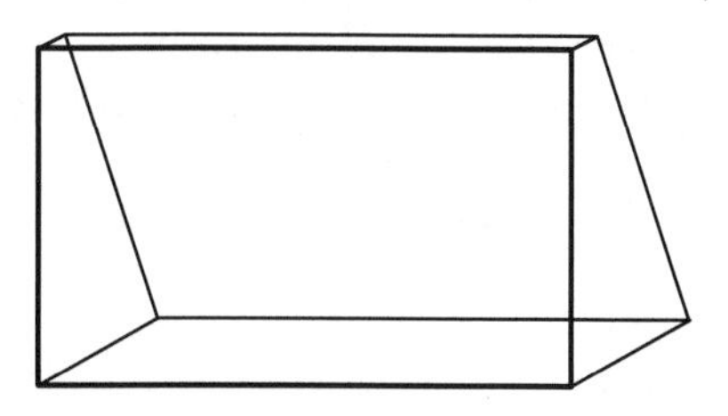

Fig. 2.1 A Goal?

This definition is irrelevant to our needs, although the GSN symbol for a Goal is indeed a rectangle. Another definition of goal is: An outcome that one strives to attain.

This is closer to the mark, and was the original meaning when the notation was proposed for representation of engineering requirements. When using GSN to represent an argument, we strive to attain the goal of persuading the reader of something, but the GSN Goal in an argument is not that type of goal either. This

J. Spriggs, *GSN—The Goal Structuring Notation*,
DOI: 10.1007/978-1-4471-2312-5_2,

would be a more appropriate definition: A Goal is a proposition to be established, i.e. shown to be true.

In formal logic, a proposition is a statement that can be either proved "True" or "False" (and, in some schemes, it could also be proved "Undecidable"). The guidance I provide in this book is not directed at proving anything formally with GSN, rather we will use it to persuade. You can think of the proposition as a claim that is being made and the truth of which will be demonstrated to the reader. A GSN Goal represents a claim, the truth of which is to be demonstrated by argument.

The GSN symbol for a Goal is a rectangle, which encloses the text of the Goal statement (Fig. 2.2). That text should be a succinct statement that can be either true or false; the argument will be that it is true.

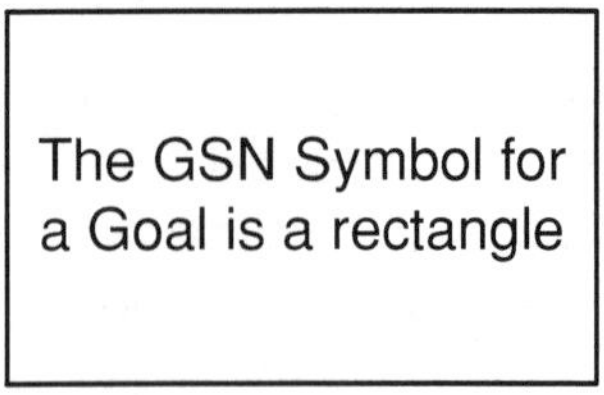

Fig. 2.2 The Goal symbol

Grammatically, the Goal text should be of the form subject-verb-object. Often the claim will be referring to a real thing, the subject, and ascribing to it an objective property. The claim must be clear. Do not use informal abbreviations, or miss out articles ("a" or "the"), and only use acronyms or initialisms if you have already established what they mean. For example, do not write, "Comm System is FFP" when you mean "The Communication System is fit for purpose".

2.2 The Top Goal: Declaring the Proposition

The overall subject of our argument, the primary claim, is called the Top Goal. There is no hidden meaning here, it is called that merely because it appears at the top of the diagram. We will construct the argument to persuade others of the truth of our claim. The purpose of a Goal Structure is to present an argument that gives the reader a high confidence that the proposition is true. It is sufficient for our arguments to be compelling, rather than incontrovertible.

I must emphasise again that you must state a Goal such that it can either be True or False; it cannot be Blue or Pulse, neither can it turn out to be 42. If it can, there is something wrong; what you have is not a Goal, not a proposition that can be argued.

It may be that you believe a claim to be true, but it is in fact false. An advantage of using GSN to prepare your argument in advance is that you are likely to find out that you are mistaken before having told the world of your conclusion…

We are not trying to prove things in GSN, merely persuade. A claim can be purely subjective, for example I could make a claim about wine grape varieties, "Syrah is better than Merlot". I may believe it to be true and make an argument on that basis, but you may believe the opposite and refute my argument. At this point bottles of wine could be opened, and we may lose track of the arguments...

2.3 Subverting a Proposition

In some circumstances you will be arguing against something; arguing that a proposition is false. If this is the case, do not "subvert the proposition"; instead, change your Goal text. Make it an opposing statement. For example, a developer wants to buy a local coppice for building houses; you want to argue against this on principle. Do not argue against the developer's statement, "This land is ideal for new housing"; rather argue for your principle: "This Ancient Woodland must not be destroyed". You can then argue that this is true, making the point along the way that to give it up for housing is, in effect, to destroy it. Of course, you would also need to demonstrate that the coppice in question is indeed ancient woodland.

2.4 How do Goals Make Arguments?

We argue for the Top Goal by decomposing it into Sub-Goals. A very simplistic example would be a Top Goal, "2012 marks the centenary of the cryptanalyst Alan Turing". This Goal can be decomposed into two Sub-Goals:

1. A person's centenary is the hundredth anniversary of his or her birth
2. Alan Turing, the cryptanalyst, was born in 1912

Once we have split the Top Goal into Sub-Goals, we then do the same to each in turn until we have a Goal Structure representing the whole argument, which would not be very large for this Turing example. Before looking in more detail at how to do that, there is something else to address in the next Chapter. It is the key ingredient that gives GSN its expressive power...

2.5 Question

- Formal logicians consider some of the compelling arguments that we will present in GSN to be "defeasible". That is, the arguments are capable of being invalidated. For example, you may attempt to persuade someone by claiming, "Well-known experts on this subject state that...". You and your readers may have confidence in those experts, whereas the logician would point out that they do not constitute the whole set of such experts, so there may be other, competing, views. Do you think that this matters when we are presenting our arguments to persuade, rather than to prove?

2.6 Problems

Which, if any, of the following are claims that could be used as a Top Goal, the basis of an argument? Can any of the others be reworded such that they can be claimed?

1. Our Quality Management System is ISO9001:2008 compliant (ISO 2008)
2. My Business Plan is complete and ready for review by the Board
3. This burial site is probably that of Rædwald, King of the East Angles
4. This painting should be attributed to Albrecht Dürer
5. This equipment fulfils the essential requirements of the RTTE Directive
6. Beryllia is a carcinogen
7. Hazard Identification and Risk Assessment
8. Assurance is provided that safety requirements raised on the software are valid
9. The GSN Symbol for a Goal is a rectangle
10. The colour of the sky

Answers to these problems can be found near the back of the book.

Reference

International Standards Organization (2008) Quality management systems. Requirements, ISO 9001

Chapter 3
Contextual Information

Abstract The truth of a claim depends on the context in which you make it. For example, if the claim is, “It is Safe”, what is it that is safe? For that matter, what is “Safe”; what do they mean by that word? You could detail the Goal statement to include all this additional information, but then it would be too cumbersome. The solution is to declare the context explicitly. This chapter introduces symbols and text conventions for declaring context; this includes the Model symbol, which is not part of the core notation, but may occasionally be encountered “in the wild”.

3.1 Background

The development of the notation that made GSN popular was for documenting safety assurance. Production of a Goal Structure would be prompted by a question from a potential user, or from the designers checking that they have done enough, or from an Industry Regulator: “Is it safe?”

The Goal statement should be succinct; in this example they may claim “The System is Safe”. There is a problem here. It may be clear to the questioners what they mean, but what about the rest of us? It just poses more questions:

- What is safe; which system do they mean when they say “The System is Safe”?
- For that matter, what is “Safe”; what do they mean by that?

As already noted, GSN is not restricted to giving safety assurance; the potential user may have asked her safety question in support of her own argument justifying the choice of a particular item for use, “Product A is better than Product B”. Again this leads to more questions:

- What are Products A and B precisely; have they been bought “off the shelf”, or have they been modified in some way?
- What criteria are to be used to assess which product is “better”?

J. Spriggs, *GSN—The Goal Structuring Notation*,
DOI: 10.1007/978-1-4471-2312-5_3,

If we were to clarify all this, the Goal statement could become too cumbersome. Remember you have to fit it all, readably, into a little box on a diagram. The following claim may be accurate, but it is impractical:

> "Version 7.2 of the Type 43-56 Protocol Conversion Subsystem, when fitted with optional modules 43-56/17288, at Version 5.9, and 43-56/29553 at Version 8.6, does not exhibit behaviour that could lead to loss of life; injuries to people or livestock; or environmental damage"

Not only would the reader say, "Sorry, what?", you would also find it tricky to maintain your argument when someone makes a change and updates a version of a plug-in module, say. That long description with version numbers probably ripples on down through the argument.

To keep our Goals succinct, and make our arguments more maintainable, we must explicitly define context for the Goal. We separate out the scope and definitions as Context.

3.2 Context

Context provides, or references out to, definitions and other supporting material. Represent it with a text box with rounded ends.

Context is always associated with Goals using an open-headed arrow. An open-headed arrow always denotes contextual information in GSN, whether it be a Context, or one of the other forms that we will meet later. Contextual reference arrows come from the sides of the Goal (or other argument element) and point into the side of the Context element symbol (Fig. 3.1).

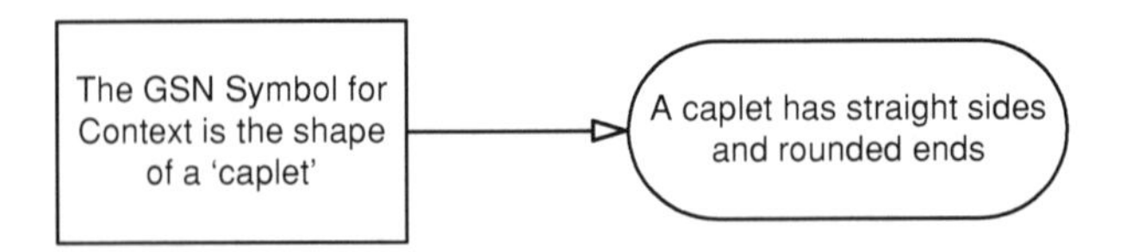

Fig. 3.1 A Goal with Context

Context symbols shall only appear at the head-end of arrows; do not hang other contextual information from them. One Context can be at the head of several arrows, i.e. providing context for more than one Goal, but check: does it make the diagram clearer, or would a repeated Context be better?

Note: The original GSN Context symbol (Kelly 1998) was not a continuously smooth curve. It was like two parallel lines with segments of a circle stuck on either end. The symbol presented above is used as a terminator in some standard flowcharting notations. It is thus available in most diagramming tools; using it is easier than making your own. Note also that some early papers (including my own) used a solid-headed arrow throughout the diagrams. This is incorrect; each contextual relationship must be an open-headed arrow.

You could get away with using a box with rounded corners for Context if there is a lot of text to fit in. We can use Context to simplify our long-winded claim of the previous page. We could, for example, reduce the Goal to, "The Protocol Converter exhibits adequate product safety" and leave it to Context to tell us what equipment constitutes the Protocol Converter, see Fig. 3.2.

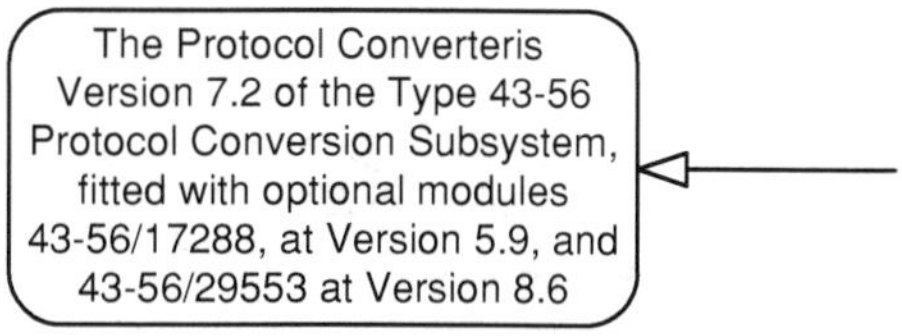

Fig. 3.2 The alternative Context symbol holds more text

The Protocol Converteris
Version 7.2 of the Type 43-56
Protocol Conversion Subsystem,
fitted with optional modules
43-56/17288, at Version 5.9, and
43-56/29553 at Version 8.6

In this case, we would also need a Context to define what properties of this Protocol Converter would constitute adequate product safety. We could perhaps also make it easier to maintain by removing the version information from the definition of the Protocol Converter and putting it into a separate, scoping, Context, "This argument is for Version…"

3.3 Context is Important

Consider an argument in which I claim my athletics team is very good, because they came second in every event. If this were in the context of the Olympic Games, it would be wonderful; we would return home to much acclaim, laden with silver. If, on the other hand, this were in the context of a match with only one other team, it would be a disaster; we would have lost everything.

3.4 Environmental Context

In a safety assurance argument, context goes further. It is usual to argue the safety of something in a particular usage environment, i.e. context. For example, as someone apparently once demonstrated, although it can be "safe" to use a vacuum cleaner to get the dirt out of the passenger compartment of a car, using it to remove petrol from the car's fuel tank can lead to a major fire.

This environmental context could be made explicit, or it could be included in the description of the thing itself. For example, a large system would include people and their procedures, which could define the operating environment; if not, the system description should have a formal concept of operations that does.

For something smaller, just what it is and where it is to be used may be enough to indicate the operating environment, as in the example of Fig. 3.3.

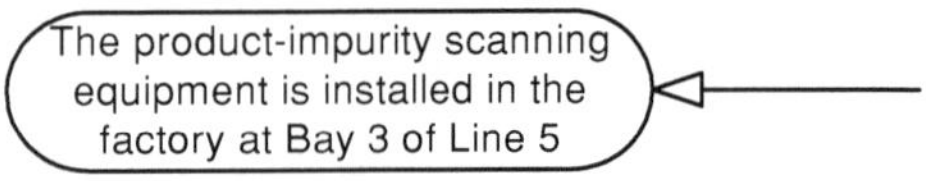

Fig. 3.3 A simple Context example

The product-impurity scanning equipment is installed in the factory at Bay 3 of Line 5

3.5 Make External References Using Context

In our earlier example, we could have provided a Context to define the "betterness" criteria needed to decide between Product A and Product B. In practice, this may be cumbersome, so I recommend that, if context is to be more than a simple clarifying sentence, it should refer out to part of an external document. If you are talking safety, it could be a definition in an external standard, or legislation; if you are talking pay rises, it could be your formal job description, or an independent report on market rates of pay.

Note that Context applies not only to the Goal to which it is associated, but also to the argument below it. So, if the argument fragment shown in Fig. 3.4 were to be expanded to consider whether particular aspects of the system are "Safe", the same definition would apply. You do not need to re-state the context, unless it is required for clarity.

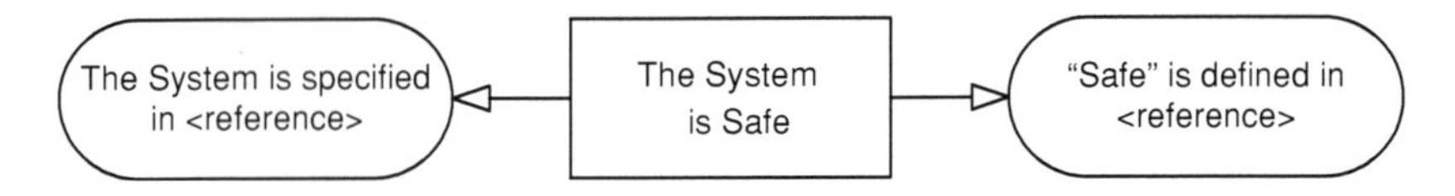

Fig. 3.4 Using Context for external references

Context lower down the argument structure may refine that higher up. The whole argument may be about the Protocol Converter, but Context can be used to emphasize that this particular bit of the argument is only about the input stage, for example. It may also widen the scope, for example, a safety argument about a device in a particular application may contain sections that are about the properties of that device in any reasonable application.

3.6 Further to Problems Set in Chapter 2

When providing answers to the problems posed in Chap. 2, I referred forward to this chapter. In the answer to Problem 5, I mentioned the ambiguity of referring to something just by its common name. The reader may wish to consult the reference, but not be able to identify it exactly; there may be several items of that name. Context overcomes this; we can be succinct in our Goal statement and use the Context to refer accurately out to the thing.

Note that, whilst Fig. 3.5, below, is correct at the time of writing, a new draft RTTE Directive is in preparation. It is likely to replace the 1999 version soon after this book is published. Note also that, in a real application, the Directive would not apply directly; it will have been transposed into national legislation with which you (or your European agent) will have to comply.

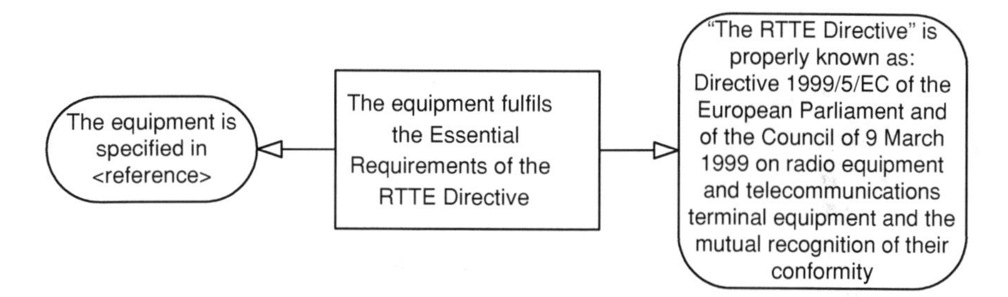

Fig. 3.5 Using Context for accurate external references

Similarly, in the answers to Problems 6 and 9 of Chap. 2, I promised a new symbol for definitions; this is, of course, Context. The example in Fig. 3.6 is that from Problem 6.

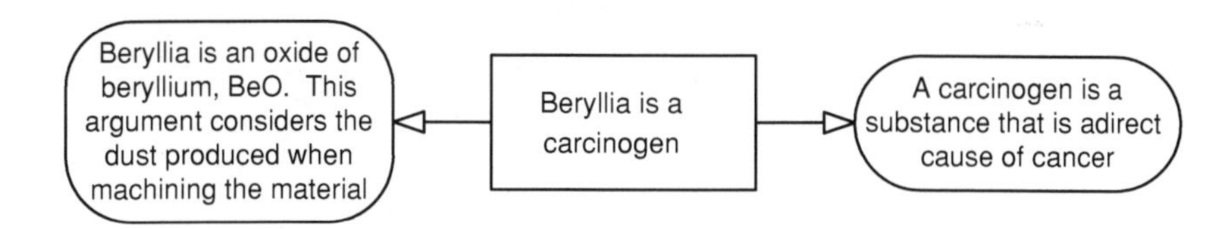

Fig. 3.6 Using Context for definitions

3.7 Avoid the Pitfall of Too Much Context

Do not take context too far. I have seen an example in which the author claims that a system is "sufficiently reliable" for some application. As well as Contexts to define the system, the application and the scope of the argument, they had two more, one to define "sufficient", the other to define "reliable". This did not clarify anything, because it still did not say how much reliability was considered to be sufficient for that particular case.

This is an important point. If your Goal statements are vague, using terms like "sufficient", "adequate", or "appropriate", it is essential to provide some criteria so that your reader will know what you are trying to achieve. If you use vague terms, your readers may expect more than you are going to give them. Their idea of what constitutes "adequate", for example, may be much more than yours. They will be disappointed; you will "lose the argument", unless you can manage those expectations at the start by giving precise targets.

The best course of action is to avoid such vague terms in your Goal statements, if you can. Clear Goal statements lead to clear arguments and reduce the need for lots of explanatory context. Conversely, too many Contexts around a Goal clutter the diagram and, hence, make your argument harder to follow (Fig. 3.7).

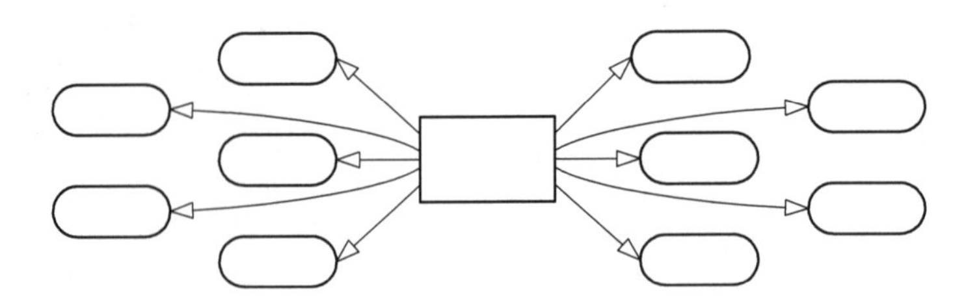

Fig. 3.7 Too much Context

Although I will present a real example like this in the answer to one of the Problems of Chap. 7, we should strive not to end up with a diagram like Fig. 3.7. It does not aid clarity, especially if each Context symbol contains a substantial amount of text.

Note that Fig. 3.7 also illustrates that the arrows do not have to be straight lines in GSN! Nor need they all come from the same point on the Goal symbol.

If you have an excessive number of Contexts because you have a lot of definitions to put across, you could collate them into a glossary document and refer out to that with a single Context. However, in most cases, it would be better to introduce the definitions just where you need them in the argument, rather than putting them all at the top to 'ripple down'.

3.8 The Model

An alternative to Context that you may encounter defining an argument's scope (particularly in an engineering environment) is the Model, which is represented with a rhombus. Note that the Model is no longer considered a core GSN symbol; most practitioners do not use it, but you may find it in published arguments.

The Model is contextual information and so, like Context, is always associated with Goals using an open-headed arrow, see Fig. 3.8.

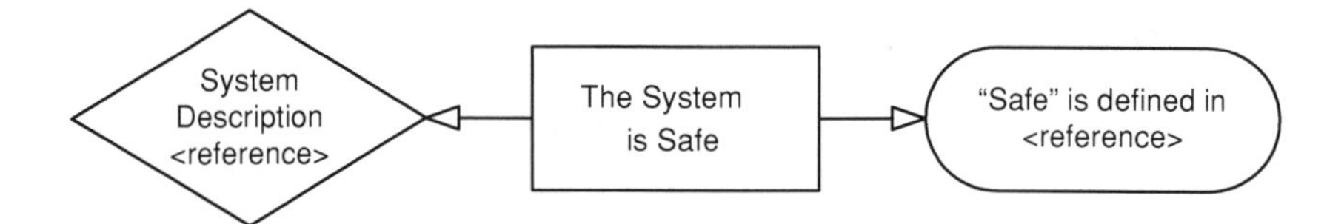

Fig. 3.8 Using a Model to specify the subject of an argument

Intended to represent the subject of the analysis that led to the argument, the Model, for example, could be a particular view from a computer-based analysis and design tool. By "view" I mean that it is some abstraction (of the system in question) that describes the aspects that the argument is about, and does so in sufficient detail to support the argument.

A Model could, for example, be a behavioural description of the subject of the argument expressed in an abstract notation, such as the Unified Modelling Language (UML 2005); it could be a simulation, it could even be a physical three-dimensional model of something. Whatever representation is used, it will be relatively large; the text in the Model symbol will always be an external reference.

3.9 Is There More Context?

I introduce other forms of contextual information in Chaps. 7 and 15. You will know when we are there, because you will see open-headed arrows associating the new symbols with Goals. Read open-headed arrows in a Goal Structure diagram as "In the Context of".

In Chap. 2, we postponed looking at how to argue for the claim of a Top Goal, by decomposing it into Sub-Goals. Now we are ready to do that, see Chap. 4.

3.10 Question

- Earlier I claimed that my athletics team is very good, because they came second in every event, and showed that the impact of this argument is critically dependent on the context. Are we Olympic heroes, going for gold next time, or are we, well, hopeless? Can you think of other examples where the same argument has the opposite effect in contrasting contexts? Note: the answer to this one will not be found elsewhere in the book.

3.11 Problems

1. My factory manager wants to persuade me that one of his preventative maintenance procedures, known as PMP5, is adequate to keep the factory power supply running. He claims, "PMP5 is fit for purpose". We want to document the argument so that in future, if someone wants to change the procedure, they will be able to ensure that they do not "break" it. Using GSN, how could the claim, "PMP5 is fit for purpose", and its Contexts be presented?
2. A fictional detective is making a case to the Crown Prosecution Service in the matter of the silver stolen from Lord Symondsbury, case reference 2011/537. He suspects that Mr Adams, one of the servants, committed the crime. His claim is, "The butler did it". If he were to present this claim in GSN, what Contexts should he provide?

References

Kelly TP (1998) Arguing safety—a systematic approach to managing safety cases (doctoral dissertation). Department of Computer Science, University of York, September 1998

International Standard ISO/IEC 19501 (2005) Information technology—Open Distributed Processing—Unified Modelling Language (UML)—Version 1.4.2, International Standards Organization and International Electrotechnical Commission, 2005 (Note that this is not the version of UML used by most practitioners, but it was, at the time of writing, the most recent formally standardized one.)

Chapter 4
How do Goals Make Arguments?

Abstract A Goal Structure is used to show how our claim, represented by a Goal, is the consequence of simpler Sub-Goals. This chapter introduces the symbol used to represent the thread of the argument and gives simple examples, showing how to construct and read simple arguments (and fallacies).

4.1 Argument

We can construct arguments by logically decomposing Goals into Sub-Goals, which are themselves Goals. You can also do it the other way around, i.e. compose an argument from some already established lower-level Goals, but that can be much trickier.

Goals are always associated with Sub-Goals using a solid-headed arrow. To follow a thread of the argument, follow the solid-headed arrows. Thread of Argument arrows come from the bottom of a Goal and point into the top of the next Goal in the structure.

One thing you may find unfamiliar at first is the sequence of argument. You were probably taught at school, when doing science reports and the like, to present your information and then draw a conclusion from it. Because of this, we are accustomed to reading documents that present a lot of information and then draw a conclusion.

The structures we are producing here do it the other way around. They first state a proposition, the Top Goal, and then show how it is supported, the Sub-Goals (Fig. 4.1). Instead of saying, "From A, B and C, I conclude D", we will have "I claim D because of A, B and C". The reasons for this are clarity of purpose and ease of reading. First state what you are going to argue about and then provide the argument.

In some arguments you may make a claim in a Sub-Goal that your readers accept already, they do not need to read your argument to be persuaded. With an argument expressed in GSN they can easily skip that section, if instead we had

J. Spriggs, *GSN—The Goal Structuring Notation*,

DOI: 10.1007/978-1-4471-2312-5_4,

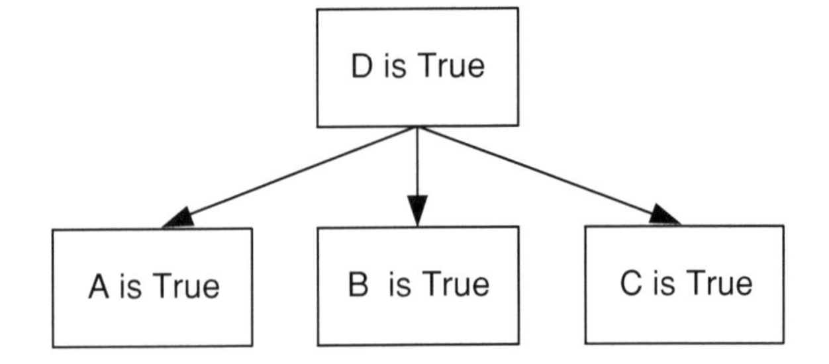

Fig. 4.1 The Top Goal decomposes to Sub-Goals

used the "present the discussion and then draw conclusions" approach, they may not have realized that they could have skipped a section until they had read it. GSN helps to focus the attention of the reader on the parts of your argument that they do not already accept.

I will present a simple argument both in text and in GSN to illustrate how the diagrams can be read. Then, to highlight a pitfall, I present a couple of fallacies; constructing an argument in GSN does not guarantee that it is valid.

4.2 A Simple Example

A simple argument from classical Greece proceeds as shown in Fig. 4.2:

Fig. 4.2 Socrates is mortal

All men are mortal
AND
Socrates is man
THEREFORE
Socrates is mortal

Figure 4.3 shows the conclusion (proposition) as the Top Goal and the premises as Sub-Goals; they are linked using solid-headed arrows.

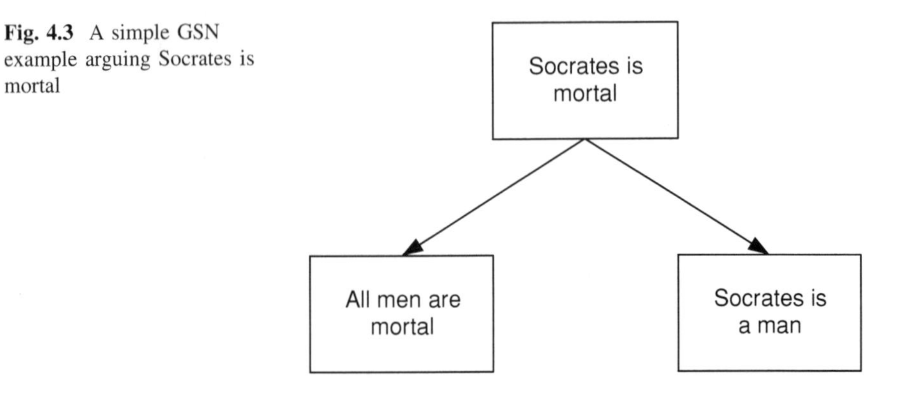

Fig. 4.3 A simple GSN example arguing Socrates is mortal

A solid-headed arrow always denotes argument in GSN, whether it be a Goal decomposition to Sub-Goals, or to one of the other argument components that we will soon get to. Thread of Argument arrows come from the bottom of a Goal (or other argument element) and point into the top of a succeeding element in the argument.

We can read the diagram as shown in Fig. 4.4:

Socrates is mortal
BECAUSE
All men are mortal
AND
Socrates is man

Fig. 4.4 Parsing the GSN 'Socrates is mortal' example

Of course a real argument would include Context; for example, who is this Socrates we are arguing about, the famous Classical Greek philosopher, or someone else? The point of the structure is that we have taken the Top Goal and logically split it into two Sub-Goals. If we can show that the Sub-Goals are true, then the Top Goal must also be true (if the logic is sound). We can carry on like this until a complicated Top Goal depends on a set of self-evident 'bottom Goals'. Of course, it is not always that easy…

4.3 Simple, But False, Arguments

GSN enables us to represent arguments; it does not constrain the argument to be correct. Woody Allen put this first example forward in "Love and Death" (Allen 1975). He, in the character of Boris Grushenko, argued, using the premises—conclusion form as shown in Fig. 4.5:

All men are mortal
AND
Aristotle is mortal
THEREFORE
All men are Aristotle

Fig. 4.5 A simple fallacy: all men are Aristotle

Compare the text of Fig. 4.5 with the GSN presentation in Fig. 4.6.

This second example (text in Fig. 4.7, GSN in Fig. 4.8) may be older. It is the fallacy called "affirming the consequent" :

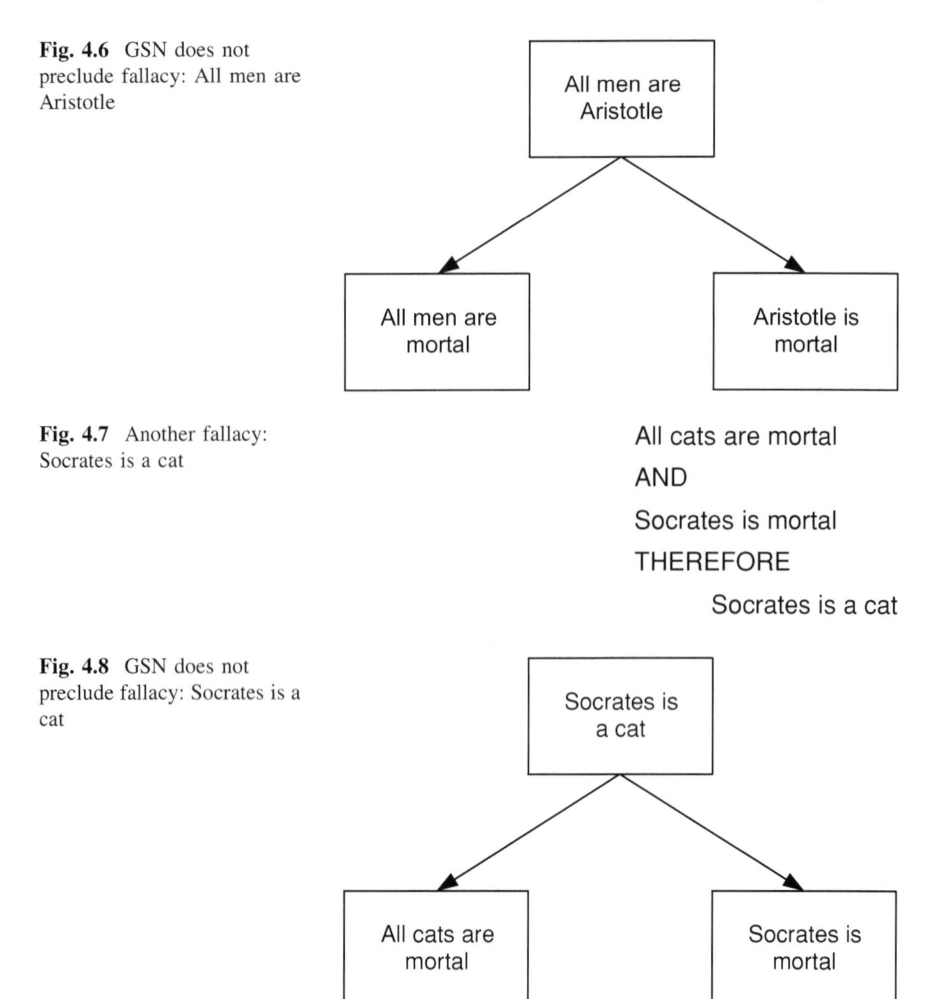
All men are Aristotle
All men are mortal
Aristotle is mortal
All cats are mortal
AND
Socrates is mortal
THEREFORE
Socrates is a cat
Socrates is a cat
All cats are mortal
Socrates is mortal

Fig. 4.6 GSN does not preclude fallacy: All men are Aristotle

Fig. 4.7 Another fallacy: Socrates is a cat

Fig. 4.8 GSN does not preclude fallacy: Socrates is a cat

4.4 Less Simple Arguments

Normally we would not argue, fallaciously or otherwise, over the mortality of people, especially if they are long dead. Also, unless we really want to make a point, we would not give a text argument as well as GSN. It is, however, worth doing at the start of a document if you are not sure if your readers are all fluent in GSN. Describe the top level breakdown in text, show them the diagram, point out that it is equivalent and then they should be able to read the rest without text descriptions. Do give a key to the diagrams that is easy to refer back to, explaining the symbols that you have used in the overall Goal Structure.

All useful arguments will have many more than two Sub-Goals in total. This next example has four at the top level, all of which will need to be decomposed further. The argument in Fig. 4.9 is saying that there is a situation that leads to some unwanted outcomes; our solution is to build a bypass and, by the way, we have already tried other ways of alleviating the situation.

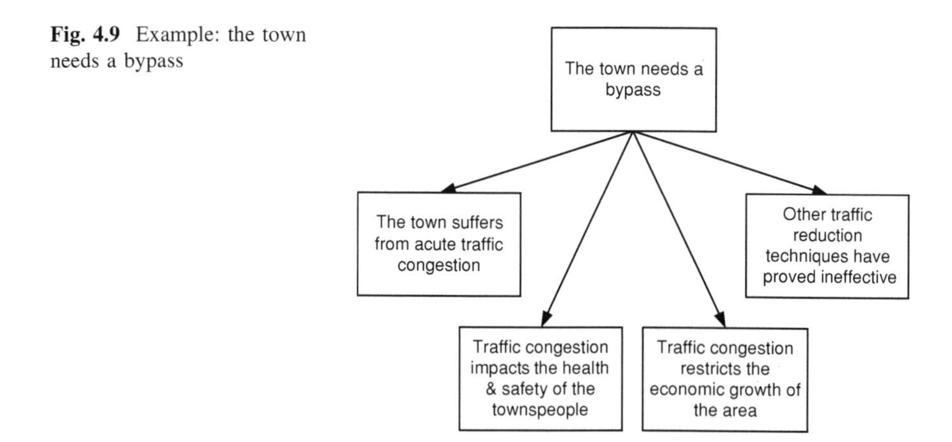

Fig. 4.9 Example: the town needs a bypass

Opponents of the bypass can prepare counter-arguments of similar complexity, or they may accept this argument, but augment it to show that having the bypass will have greater adverse economic, health and safety impacts than not having it.

I oversimplified this example just to emphasise breaking the Top Goal down to multiple Sub-Goals. A real argument would have had some contextual information, perhaps stating that a bypass would divert traffic around the town and so reduce congestion within it. This example also needs geographical detail, and not just stating which town; a north–south bypass would not help if it were predominantly east–west traffic causing the congestion…

An alternative to a Context stating the purpose of a bypass would be to have an extra Sub-Goal, "A bypass will reduce traffic congestion in the town". The choice really comes down to whether you expect it to be self-evident to the reader. If you are making your argument to the local government transport committee, they should know what a bypass is and how it works; your Context reassures them that you have the same understanding. If, however, you are trying to persuade the pupils of the local primary school, you may need additional argument to explain the benefits.

4.5 Reading Order

Note that, in Fig. 4.9, I arranged the Sub-Goals in a logical order reading from left to right. This was just an aid to comprehension; it is not a requirement of the notation. Usually the Sub-Goals will be independent things; they can be taken in any order. This is what the notation expects but, if you can make your argument clearer by ordering, it will be to your advantage to do so. Formally, the property allowing the premises of the argument to be stated in any order without changing the meaning is called "commutativity of conjunction"; you do not need to remember that name to use GSN…

An example of an argument whose Sub-Goals can be considered essentially independent, i.e. presentable in any order without loss of clarity, comes from fiction. I will first present it in text form (Fig. 4.10) and then in GSN (Fig. 4.11) with the Sub-Goals (premises) in a different order. It is the same argument, regardless:

Fig. 4.10 Example: the defendant is guilty of the crime

The Defendant is Guilty

BECAUSE

He had the means to commit the crime

AND

He had a motive to commit the crime

AND

He had the opportunity to commit the crime

This seems to be sufficient in many TV dramas; in practice you would need a lot more. In particular, you would have to argue that this defendant actually is the person who took the opportunity to commit the crime, using the means presented (and then you may not need to explore motive). You would also have to argue other things like 'competency' or 'criminal responsibility'.

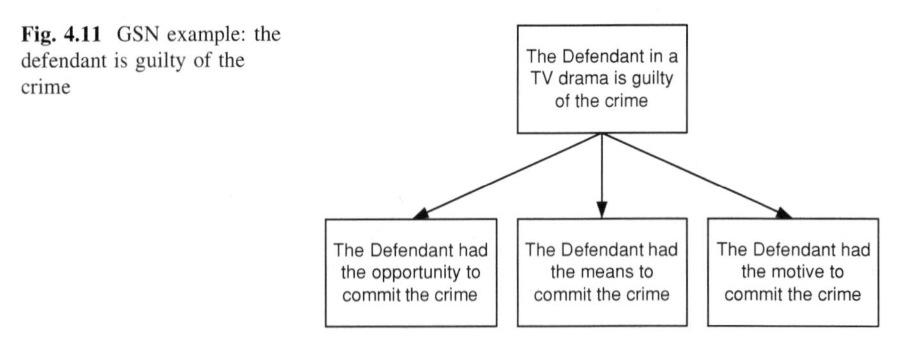

Fig. 4.11 GSN example: the defendant is guilty of the crime

4.6 Challenging the Argument

We will use GSN to make our arguments clear, easy to read and, hence, easy to challenge.

When you develop an argument in GSN, you should challenge yourself as you go along. It is much easier to make the argument stronger by fixing, or removing, any weak points in small sub-arguments of the size of the examples in this chapter than it would be to examine the overall Goal Structure in one go.

Similarly, if you are reviewing an argument, look at the logic of each level in isolation; are there any holes?

I will come back to these points in Chap. 9.

4.7 Can It be Clearer?

The arguments presented have been a bit "dry" and leave the reader to do a lot of work to gain understanding. Here is another, written in a similar style, arguing that a mobile telephone base-station is secure (sufficient to deter people from breaking in to steal equipment). The actual level of security required is defined by the company's Security Management Manual, which also specifies requirements for warning signs. Figure 4.12 presents a slightly obtuse version; I then offer an alternative, and hopefully clearer, version for comparison.

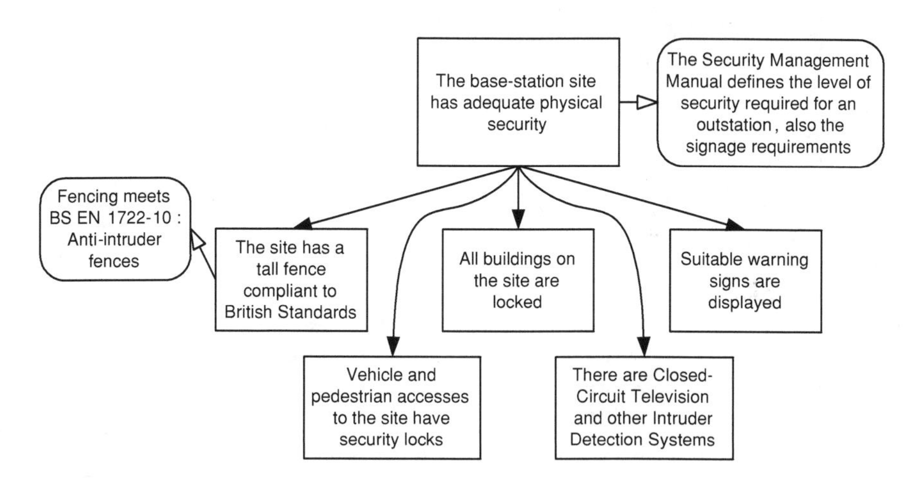

Fig. 4.12 Security example: the barebones version

Now compare that with the following version (Fig. 4.13). Which of the two is more compelling, because the argument is more obvious to the reader?

Most audiences will prefer this explanatory style over the dry logical form. There are other ways of explaining the rationale of an argument without cluttering up the Goals. One is a form of contextual information, which will be introduced in Chap. 7; the other is a new argument element, see Chap. 6.

All the argument examples so far have been superficial and not particularly useful, other than to introduce the symbols. What about something more practical? In Chap. 5, I use the notation that we have met so far in a (fictitious) business

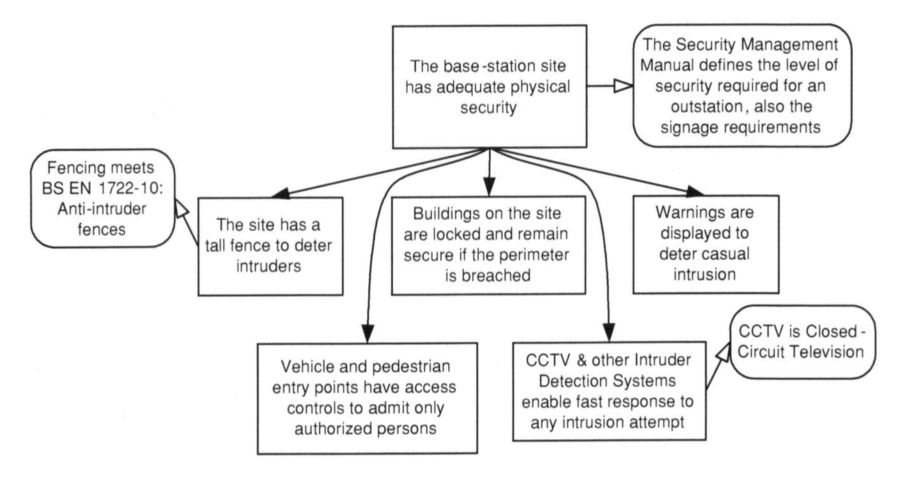

Fig. 4.13 Security example: an improved version

application. The Goal Structure will be more complex, so I will address how to present it to your audience; how it can remain readable even on small pages.

4.8 Questions

- If, like me, you were bought up on Boolean algebra and propositional logic, you may have looked at Fig. 4.1, which says A, B, C, therefore D and thought it invalid. That is because here we are using a form of predicate logic. Underlying Fig. 4.3 there is a specification: For all {Name}, {Name} is a man implies {Name} is mortal. In this case: Socrates is a man, so Socrates is mortal. Do we need to specify the type of logic that we are using in our Goal Structures, with a Context on the Top Goal for example?
- Did you notice that, whilst the first example given (the one about Socrates being mortal) could be checked by simply working through the logic, the later examples are not like that. The Socrates one is in some way complete: all men are mortal; the set of all men contains Socrates, so he must be mortal. The later examples are less 'concrete'. For example, I could have argued for a bypass without considering economic growth and no-one would have said that the logic does not work. This is why I say that we will be using GSN for argument, not proof. But what is happening here, why is my reasoning acceptable even if it is not apparently logically rigorous (and are reasoning and argument the same thing anyway)? Note: An argument that proceeds without any room for probability is called a "deductive argument", whereas one that is based upon the estimation of the probable truth of the premises is called an "inductive argument" (Cass 1993).

4.9 Problems

1. In an early Science lesson at school, you were told that litmus paper goes red in acid and blue in alkali. You then observed that litmus paper goes red in vinegar. Express your conclusion (and these premises) in GSN.
2. Fletchings are the fins attached near the back of an archery arrow. If it had no fletchings, an arrow would be subject to uncontrolled rotation about its centre of mass when in flight; this is called tumbling. Any object would do this, it is not peculiar to arrows. If fletchings are fitted, they would be subject to aerodynamic forces if the arrow were to deviate from flying arrowhead first. These forces act so as to make it fly point first. Construct a simple argument in GSN claiming that Fletched arrows fly without tumbling.
3. Adelle is the daughter of Bertrand and Celine. Celine's parents are Didier and Estelle. Construct a simple argument in GSN claiming that Adelle is Estelle's granddaughter. Note that, in Chap. 5, I will use the answer of this problem as the basis of a (slightly) more complex one.

References

Woody Allen (writer, director and star) (1975) Love and death. United Artists film
Cass M, Le Poidevin R (1993) A logic primer. Vortext Publishing Ltd

Chapter 5
A More Practical Example

Abstract We now have enough notation to do something practical. An example is presented arguing that a (hypothetical) departmental business plan is complete, and showing how such an argument can be used as a tool that any department in the company can use to "dumb down" preparation of their departmental business plan without loss of confidence in the results. This example introduces ideas for partitioning large arguments for publication, for unambiguous labelling of parts of the argument, and for initial roughing-out of the drawings

5.1 The Problem

We have enough notation (Goals, Contexts, and two sorts of relationship arrow), to do something practical. This is a business example, but the principles can be applied much more widely.

Larger companies tend to be split up into divisions, in turn sub-divided into departments. Each of these should have defined rôles and responsibilities and will be set some objectives to achieve over the next operating period. They will also have some locally generated objectives; for example, to improve the way they operate so as to reduce waste, or they may need to replace some equipment that is no longer maintained by the manufacturer. All this is captured in a department business plan showing what is to be achieved by when. Such a plan tends not to say how, but it will say by whom (and it is up to them to work out the how).

If I were the department manager; how would I know when my plan was ready for formal issue and presentation to the Board of the company for approval?

Remember, when we were talking about contextual information, I said that development of a Goal Structure could be prompted by designers asking if they have done enough? It is the same here. I need to recast my question as an affirmation, "I have done enough!" and argue to show that it is true.

I grab a sheet of paper and quickly rough-out the top levels of the argument, "My Business Plan is ready for review by the Board". This Goal may seem familiar; its claim was stated in the answer to Problem 2 of Chap. 2.

J. Spriggs, *GSN—The Goal Structuring Notation*,
DOI: 10.1007/978-1-4471-2312-5_5,

5.2 First Cut

The first draft really was rough, so I will not reproduce it here. It was sufficient to show where this could lead; it can do more than allay fears about the Board.

I wanted a presentable copy to show people, so I prepared the more formal version shown in Fig. 5.1. It is not just about my business plan now; I have spotted

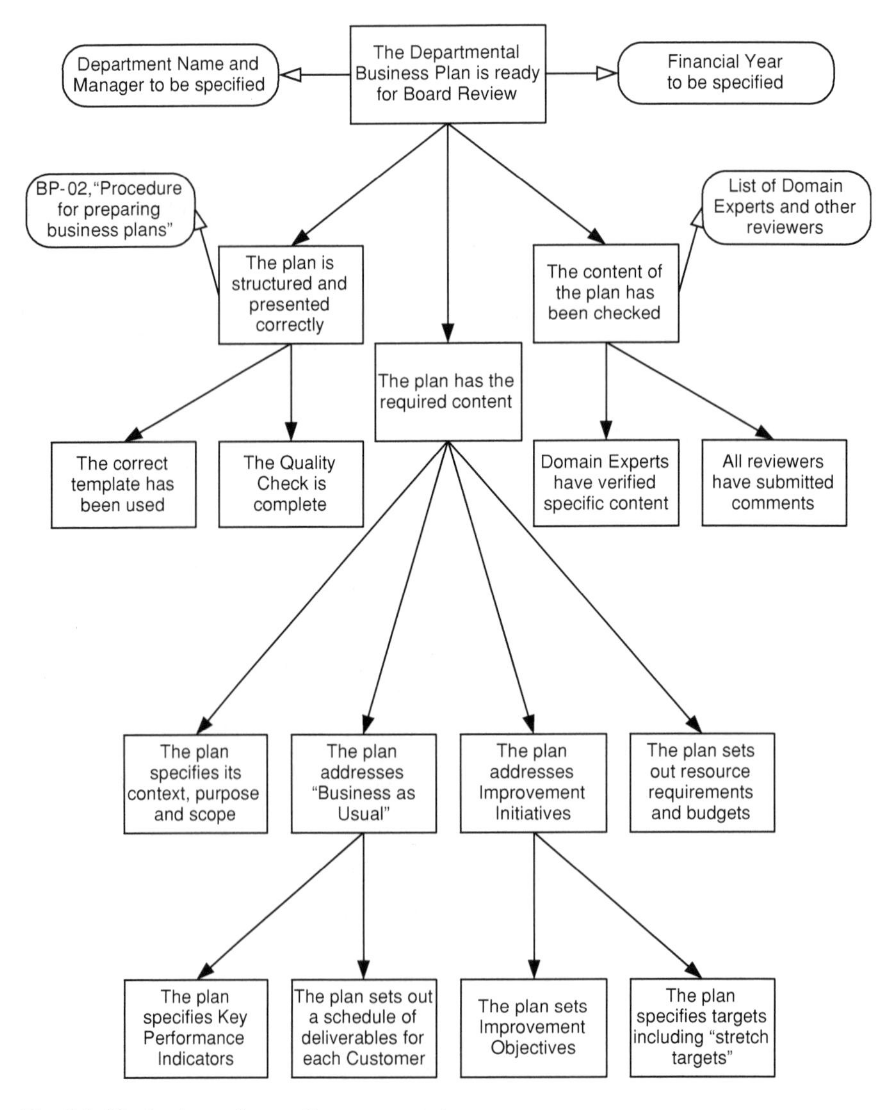

Fig. 5.1 The business plan readiness argument

that my argument could be of use to other departments too. The Top Goal could thus be "The department's Business Plan for next year is ready for Board Review", but I'll slip the year into Context so that it can be stated explicitly.

At this stage, I need to make a number of observations. They are actually independent of the subject matter of the example, but we will come back to that soon and use this Goal Structure (Fig. 5.1) as the basis of a useful management tool.

5.3 Segmenting a Goal Structure for Publication

The first observation is that it is getting a bit large, especially for a small format book; if we continue like this, the text will become too small to be readable. We can overcome this problem by splitting the argument up for publication. Do not show the whole thing on one page, instead present sub-structures.

For example, I have presented the top half of the previous argument as Fig. 5.2, with the Sub-Goals of "The plan has the required content" not displayed thereon. I have repeated that Goal on the next page, but this time with its sub-Goals (Fig. 5.3).

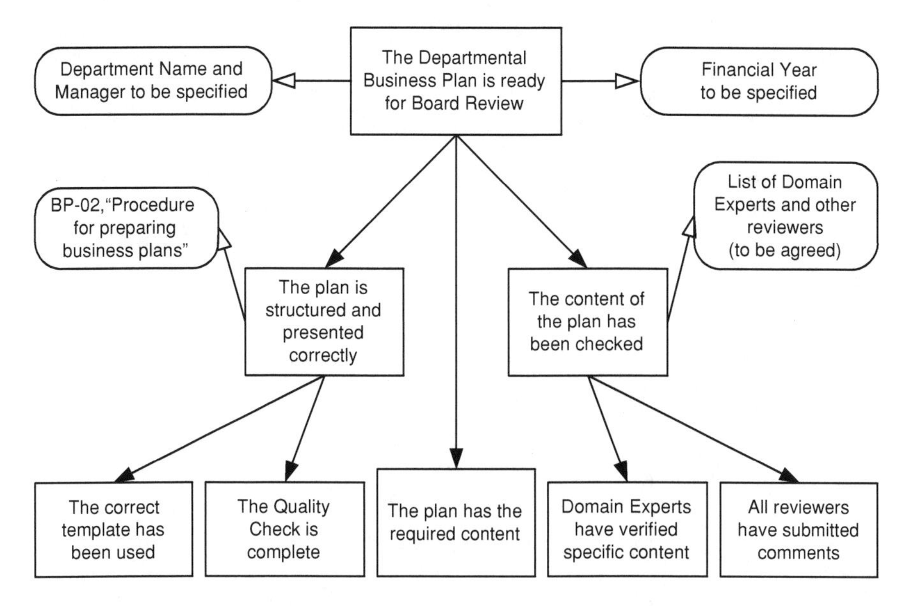

Fig. 5.2 The business plan readiness argument, Top Goals

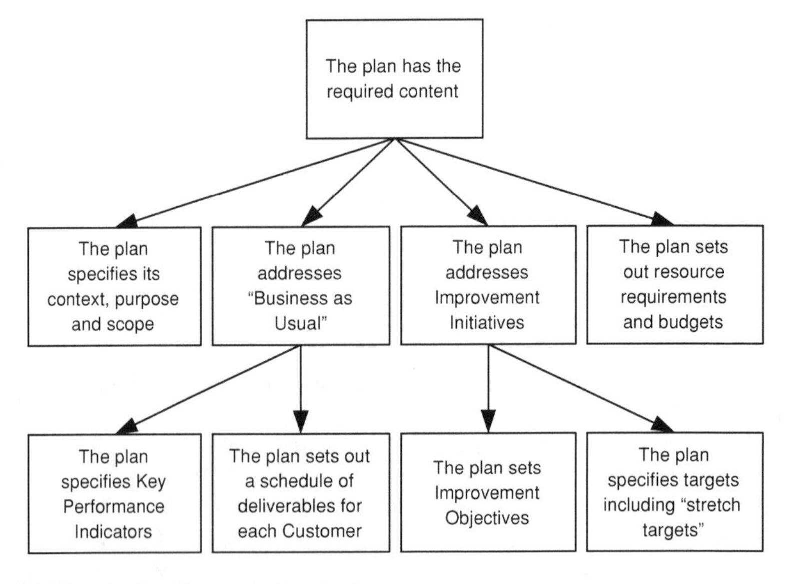

Fig. 5.3 The plan has the required content

5.4 Labelling a Goal Structure for Reference

The second observation is that, when making the first just then, I had to refer to a Goal by its proposition, "The plan has the required content". Fortunately it was a short one. In practice, if you want your arguments discussed or reviewed, unambiguously label all parts of the structure, i.e. all Goals, Contexts, etc. There are a number of strategies for doing this. For ease of navigation I prefer a hierarchical scheme; I label the Top Goal G0 and its Sub-Goals are then G1, G2... The Sub-Goals of G1 are G1.1, G1.2... Then there is G1.1.1, G1.1.2, etc., below G1.1, and so on (Fig. 5.4). I put the label inside the symbol, but I have seen examples where

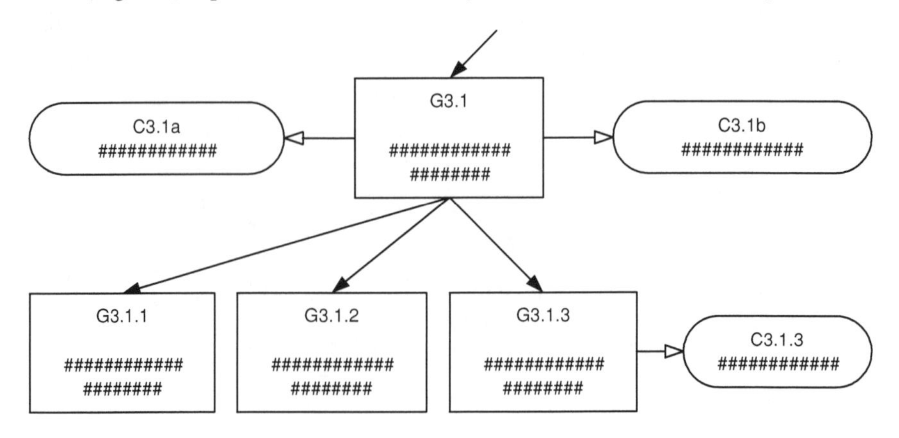

Fig. 5.4 Labelling on a Goal Structure

it is placed on the diagram near the symbol to which it relates. This latter approach gives more room for text, but it could get messy if you need to reposition the symbols on your diagram.

Having done this, I can look at a Goal label, when I am way down in the detail, to see both which top level Sub-Goal it ultimately supports and how far down the structure I am. This scheme also has an advantage when some low-level Sub-Goals are repeated. One of the examples in Chap. 13 illustrates this; the same Sub-Goal appears in four places as it supports four Goals, which are shown in separate Figures. In most cases, it would only be necessary to decompose a repeated Goal the first time, because the second time it appears, we will be able to tell from its label that it is documented elsewhere. By all means, show its decomposition in both places, if that will lead to greater clarity, but make sure that it is decomposed the same in both places and that the decomposition will stay the same when the argument is maintained by others.

Other labelling schemes I have seen used are:

- To number each Goal sequentially so, if there are thirty Goals in total, they will be numbered from G1 to G30
- To assign a mnemonic label to each Goal, e.g. ReqCont for "The plan has the required content". This is quite difficult to sustain in a large Goal Structure as the names must be unique. (This scheme is used in Chap. 15).
- To number each Goal in a manner that is seemingly random, but relates to its position in an underlying data structure. I have seen two instances of this; it was not as inconvenient as it sounds, as the data structures in question were, in both cases, linked to graphical representations that could be searched electronically in computer-based tools.

There is also no consensus as to how you should number Contexts, all the bullets above apply. I think it is clearer if you include the identity of the Goal that the Context supports as part of the label, for example C1.1, indicates that this is the Context that goes with Goal G1.1. Of course, this system breaks down when there is more than one Context per Goal, so you could use a composite form like C1-G1.1, C2-G1.1. I prefer the form C1.1a, C1.1b, which is more compact, see Fig. 5.4.

5.5 Layout and Balance

My third observation is that the Goal Structure is symmetrical. This is not a requirement of the notation; it just looks nice. That said, Fig. 5.1 just turned out to be symmetrical, I did not force it to be so (well, other than moving the 'list of reviewers' Context up a level, see Problem 1 below).

The structure will reflect the shape of your argument. Do not contrive it to look nice. However, if the shape is very lop-sided, i.e. some high level Goals have lots of levels of Sub-Goals, whereas others have very few, it may mean that your original breakdown was not really appropriate. It can also mean that you are glossing over some part of the argument of which you are unsure.

If you are a reviewer of a lop-sided argument, look carefully at the sparse bits. Alternatively, are they trying to hide something in the complex bits, so that you cannot see the wood for the argument trees?

5.6 Strategies for Drafting a Goal Structure

Lastly, and this is not so much an observation as a confession. When I said "I grab a sheet of paper and quickly rough-out the top levels of the argument", I actually used my A5 sized sketch pad, which was a poor choice for drafting a larger Goal Structure… I had to start again on a bigger piece of paper and that turned out to be a bit cramped too. In practice, a pad of "yellow stickies" is probably the best place to start.

Write your initial set of Goals on stickies, one per Goal, and put them on a whiteboard, or similar, so that you can draw (and readily redraw) the arrows to develop the structure.

You can use stickies of a different colour for contextual information, or just quickly sketch a symbol in the corner so you can see what it is meant to be, see Fig. 5.5 below.

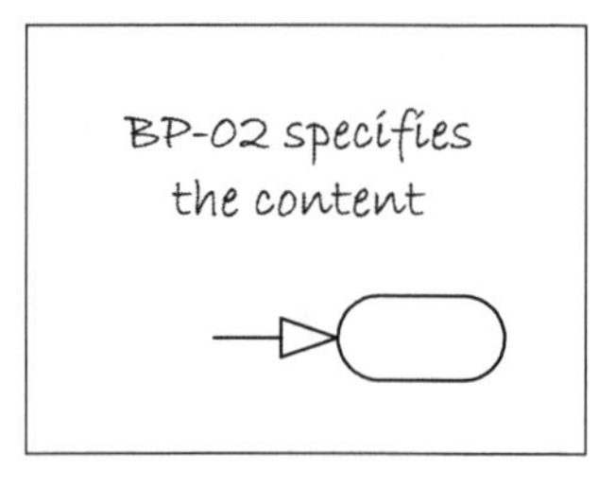

Fig. 5.5 Sticky Context

I have also seen Goal Structures drafted using index cards pinned to a cork-board, with bits of 'red tape' ribbon for the relationships. Believe me; it is easier with a whiteboard and stickies…

Chapter 6 will also consider strategies for drafting a Goal Structure.

5.7 The Business Plan is Ready

Having looked at presentation and development, we can now return to the content of the example and note that, in this case, it is a significant benefit to be generic. A manager would probably start her argument development by explicitly stating the department name, the financial year in question, etc. Some of the Goals would be very specific and informal, "Ed reviewed the figures". Once she realises that her argument has wider application, she should go back and make it more generic, "An accountant has reviewed the budget".

When she has finished she will have a tool that any department in the company can use to "dumb down" preparation of their departmental business plan without loss of confidence in the results. In fact the results should be better than when each department manager approached it in an ad hoc manner. How can this be?

Remember that the Goal Structure has been developed such that the truth of each Goal follows from the truth of each of its set of Sub-Goals. If the lowest-level Goals are true then the top one must be, if the argument is sound. See Fig. 5.6, which argues about argument.

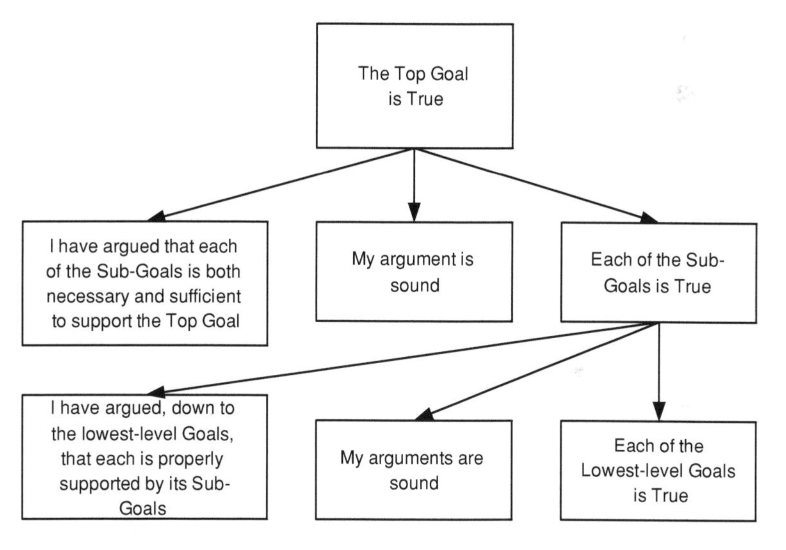

Fig. 5.6 Arguing about argument

At the bottom level of the business plan argument, we will have Goals like, "An accountant has reviewed the budget", "Each objective owner has accepted the resource constraints", "Each objective owner has agreed that their objective is achievable in the timescale", and so on. We started with a question turned into a Goal statement; we can now transform our lowest-level Sub-Goal statements into questions on a checklist; each can then be ticked-off when it is done.

We have gone from an anxious manager wondering if her plan is good enough, to a potential high-flyer proposing a means of verifying that everyone's plans are fit for purpose and ready for issue to the Board.

The business plan argument exemplifies an application of GSN that may not have previously occurred to you. Produce a comprehensive argument that some task is complete, and then use the lowest-level Sub-Goals as the basis of a checklist that people can use to verify completeness when they do the same task. The people doing the work and applying the checklists do not have to be familiar with GSN as either readers or writers. The GSN would be used to persuade the responsible managers that the checklists ask the right questions and that it is complete. In a business context, Quality Assurance practitioners, say, could develop arguments and checklists that the producers can then use to assure the quality of their work.

As well as constructing checklists from scratch, we can use an argument to verify an existing one. Construct and refine the argument down to the level from which you can make a checklist. Compare this list with the current version. The questions will not be the same, but you should be able to identify equivalences. Mark equivalent pairs with a highlighting pen. What is left? The unhighlighted parts of your newly generated list, if any, should be added to the old one; these are the questions that were missed out originally.

What about unhighlighted parts of the original list? There are two reasons why you may have such questions left over. Either your new argument is incomplete, or the checklist is asking something unnecessary. If the former, go back and modify your argument to encompass the bits you missed, generate a new checklist and repeat the highlighting exercise. If you are really sure that your argument is correct and complete, delete the unnecessary material from the original checklist.

5.8 Questions

- We build arguments in GSN by decomposing our claims into (ostensibly) simpler ones and illustrating the relationships. Is that good enough? We are trying to persuade our audience of something. Is it always possible to find a decomposition that is obvious to the reader; that the reader will see to be correct? In particular, real examples will not always be simple splits into two or three Sub-Goals. Even if you were to use the more explanatory style shown at the end of Chap. 4, could you always make a seven-way split, say, convincing? Chap. 6 presents a way of explaining the rationale of an argument without cluttering up the Goals.
- Is there anything in your work that would benefit from a checklist made from an argument? If so, try and make one!
- Similarly, do you use a checklist that can be verified with an argument? If so, go and verify it!
- Would it be feasible to produce an argument, and hence a checklist, to answer the question, “Is my argument, presented in GSN, ready for use?”

5.9 Problems

1. Do you recall the family of Chap. 4? Adelle is the daughter of Bertrand and Celine. Celine's parents are Didier and Estelle. Estelle's brother Frederic is married to Gabrielle; they have a son, Henri. Construct an argument in GSN that Gabrielle is the great aunt of Adelle by decomposing this Top Goal into two appropriate Sub-Goals, and then decompose your Sub-Goals one level, as necessary.
2. Construct another argument in GSN that Gabrielle is the great aunt of Adelle by decomposing this Top Goal into appropriate Sub-Goals, one of which is the answer from Chap. 4, "Adelle is Estelle's granddaughter". Decompose your other Sub-Goals one level, as necessary.
3. My original draft of Fig. 5.1 had the Contexts for domain experts and reviewers shared, as in Fig. 5.7, below. Why was I able to move it to the higher Goal without changing the meaning?

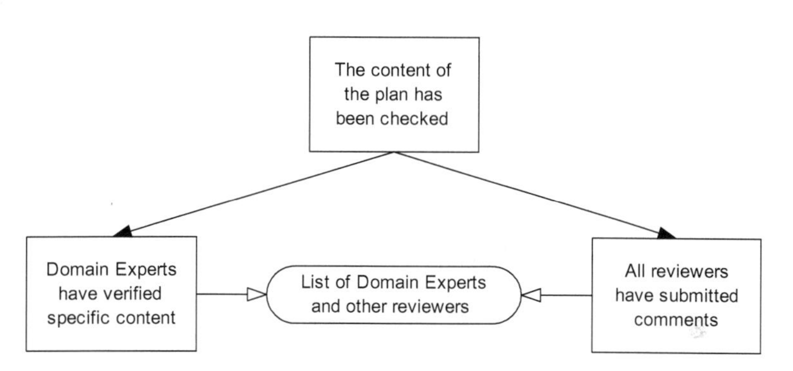

Fig. 5.7 Original location of the Context

Chapter 6
Argument Strategy

Abstract This chapter introduces the concept of Argument Strategy, which is used to express the rationale behind choice of Sub-Goals. It describes the symbol and text conventions used to represent Strategy in Goal Structures. Having now introduced the basic argumentation symbols, we are in a position to construct some serious arguments. This chapter also presents a process flow, formalising how this can be done by developing claims and strategies.

6.1 Explaining Your Choice of Sub-Goals

Sometimes, when splitting a Goal into Sub-Goals, the rationale behind choice of Sub-Goals will be obvious to the reader, but a more complex, or obtuse, argument will require explanation.

You could provide your rationale in some accompanying text, but that would defeat the object of using GSN in the first place. An argument that requires the reader to go off and read explanatory text is not a clear argument.

GSN has a symbol that you can use to present your rationale on the diagram. It is called Strategy.

6.2 Declaring Strategy

Strategy is used to express how the Sub-Goals address the parent Goal. Represent it with a parallelogram containing the strategy text, see Fig. 6.1.

Strategy is typically phrased as "Argument by...", "Argument over...", or "Argue that..." although other forms may be used. For example, you could make it like a label, such as "Completeness Argument", or just a statement of strategy, like "Show that the symptoms presented match the diagnostic criteria".

J. Spriggs, *GSN—The Goal Structuring Notation*,
DOI: 10.1007/978-1-4471-2312-5_6,

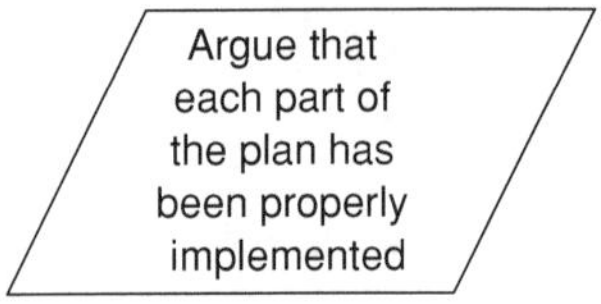

Fig. 6.1 The Strategy symbol

Argue that each part of the plan has been properly implemented

The example in Fig. 6.1 uses the "Argue that" form; it comes from a task completeness argument wherein a Goal has been stated that a plan has been properly implemented, but the Sub-Goals have been expressed in terms of individual parts of the plan, e.g. claiming completion of Hazard Identification, Risk Assessment, Design Mitigations and so on. Without the Strategy it may not be clear to a non-technical reader (or someone who has not read the plan in question) that this is the basis of the decomposition into Sub-Goals.

6.3 Strategy in the Goal Structure

The Strategy is part of the argument and so always links to Goals using solid-headed "Thread of Argument" arrows. It will always have only one incoming arrow from above, and many outgoing arrows below. Contextual information may be referenced with open-headed arrows from the sides of a Strategy in just the same way as for a Goal.

A Strategy pertains to the Goal immediately above it in the structure, so if you are using my labelling scheme (see Chap. 5), give it the same label as the Goal, but with an S prefix to indicate Strategy. For example, I would label the Strategy for Goal G1.1 as S1.1.

When we had two Contexts to Goal G1.1, I suggested labelling them as C1.1a and C1.1b. However, if you were to invoke two Strategies, which is a valid, but relatively rare, structure I suggest that, instead of S1.1a and S1.1b, you use S1.1.1 and S1.1.2 (Fig. 6.2). This is because they are part of the argument and so should be reflected in the hierarchy; it also avoids confusion by keeping letters out of the hierarchical numbering down the argument tree.

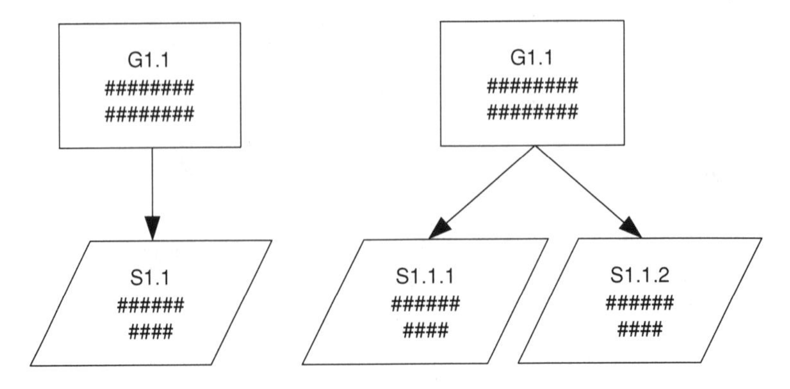

Fig. 6.2 Labelling Strategy

In this latter case, a Context attached to the first Strategy can be labelled C1.1.1 uniquely. If, however, a Context is attached to the single Strategy S1.1.1 below a Goal G1.1.1 that already has a Context C1.1.1, it has to be labelled differently. I suggest using the C1.1.1a and C1.1.1b convention, using letters here would not cause confusion, because a Strategy is intimately tied in with its Goal (Fig. 6.3).

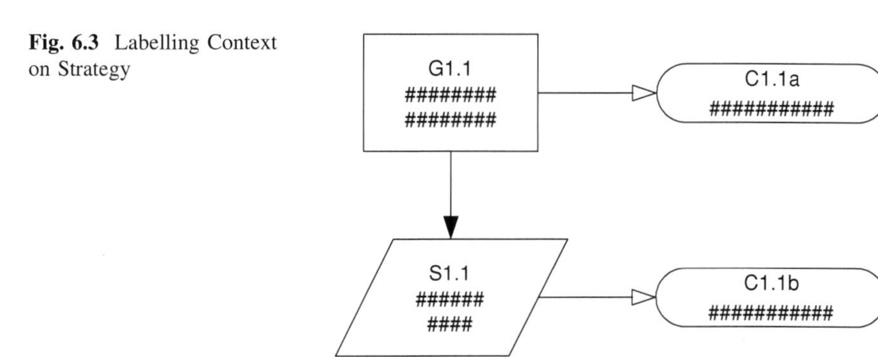

Fig. 6.3 Labelling Context on Strategy

A Top Goal, G0, may have two Strategies, S1 and S2, as shown in the example below (Fig. 6.4).

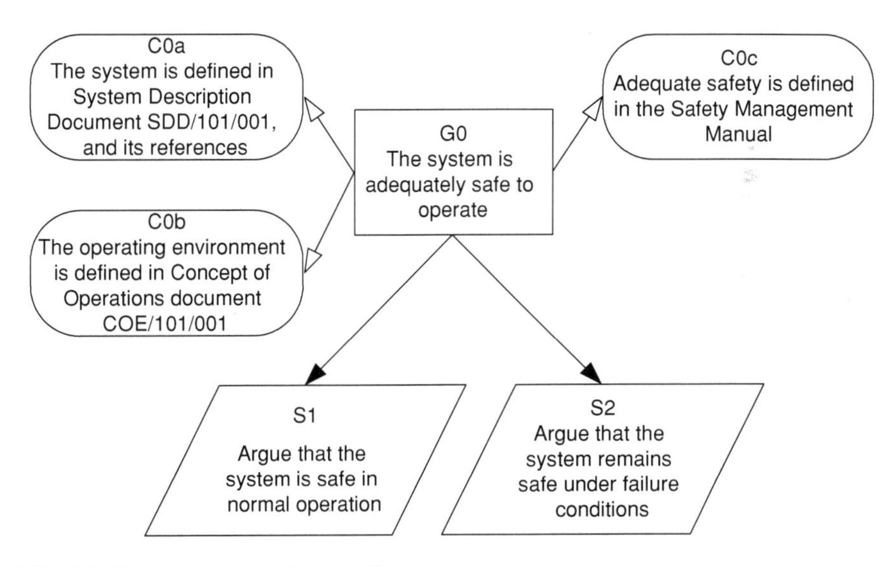

Fig. 6.4 You can use more than one Strategy

In general, this sort of structure arises when you have two things about which you have to argue, but your Customer has asked for a single argument. It may also be used when your Top Goal has two parts: "The system is safe to put into operation now, and shall remain safe in long-term operation", for example.

There may be cases where you want different aspects of the problem to be argued by different specialists or external suppliers; you could make the partitioning explicit with multiple Strategy, as shown in Fig. 6.5, although it is more usual to see Sub-Goals (or a new type of contextual information to be introduced in Chap. 7) used for this.

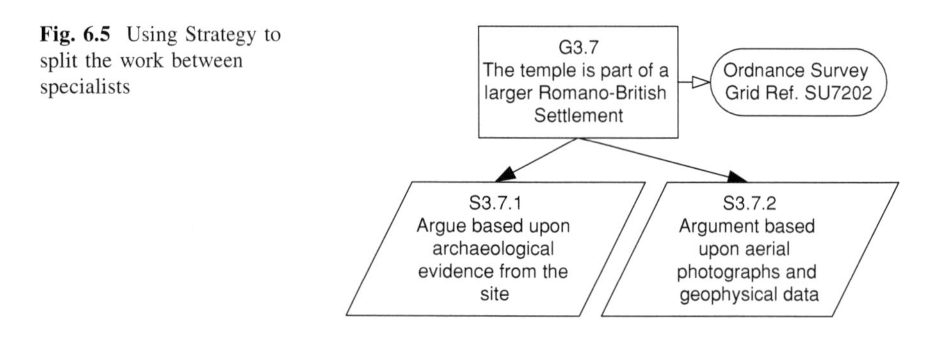

Fig. 6.5 Using Strategy to split the work between specialists

6.4 Using Strategy to Emphasise a Point

You need not restrict Strategy to complex arguments. We can also use it where the argument is simple, but we want to emphasise a point to the reader. In the example of Fig. 6.6, I am arguing for the selection of Option C, received in response to an Invitation to Tender, but have emphasised with the Strategy, and its Context, that I assessed the whole set of candidates; I did not arbitrarily ignore some of them.

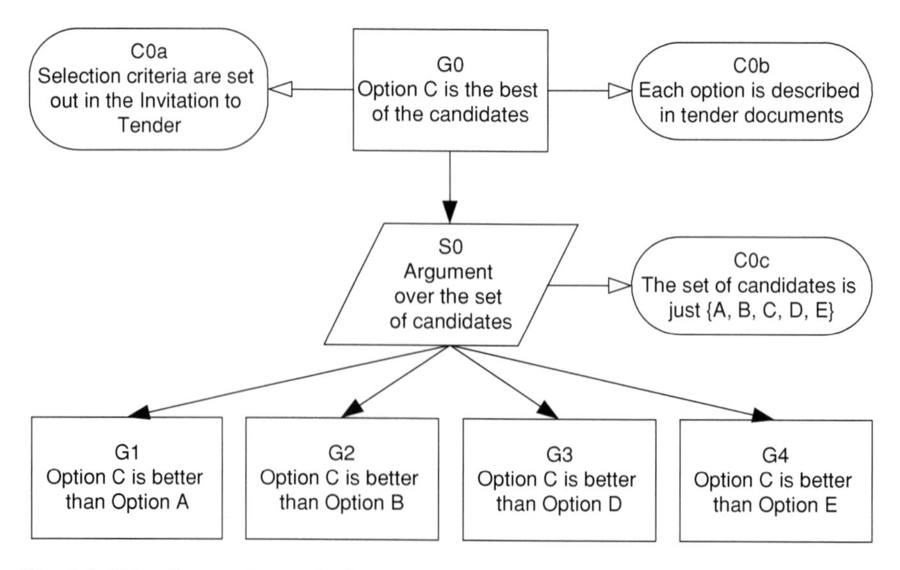

Fig. 6.6 Using Strategy for emphasis

Of course, if you were really selecting an option to spend your money on, you would also want to know if Option C is a valid choice. We may have the fallacy of "damning the alternatives". C may be the best, but it may still be inadequate; it just happens to meet more of your requirements than the others. At the other extreme, C may be best, but B is sufficiently good and it is cheaper, so you would buy that instead.

Again, I've oversimplified, I have not claimed, for example, that if C is better than all the others then C is the best. This may seem needlessly fussy, but with some audiences it is necessary to point out the seemingly obvious. There is a symbol to facilitate this, which we will meet in Chap. 7.

Using a pair of Strategies could also be thought of as a form of emphasis. In Fig. 6.4, I emphasise that the argument is making two distinct points; it will be safe in normal operation, but if something were to go wrong, it would be safe then too.

6.5 Further to Chapter 4

Chapter 4 spoke of making arguments easier to understand by augmenting the Sub-Goal text with a bit of explanation, an alternative to this is to use Strategy to carry the explanation. The choice depends on which you think will be clearer for your reader. Figure 6.7, below, is the site security example from Chap. 4, Fig. 6.8 makes the same argument using Strategy to simplify some of the goals.

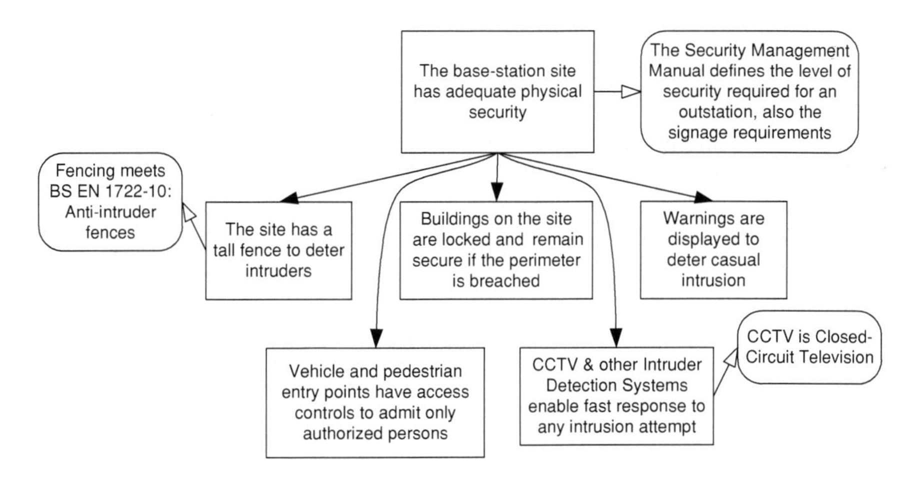

Fig. 6.7 Security example: the "improved version" of Chapter 4

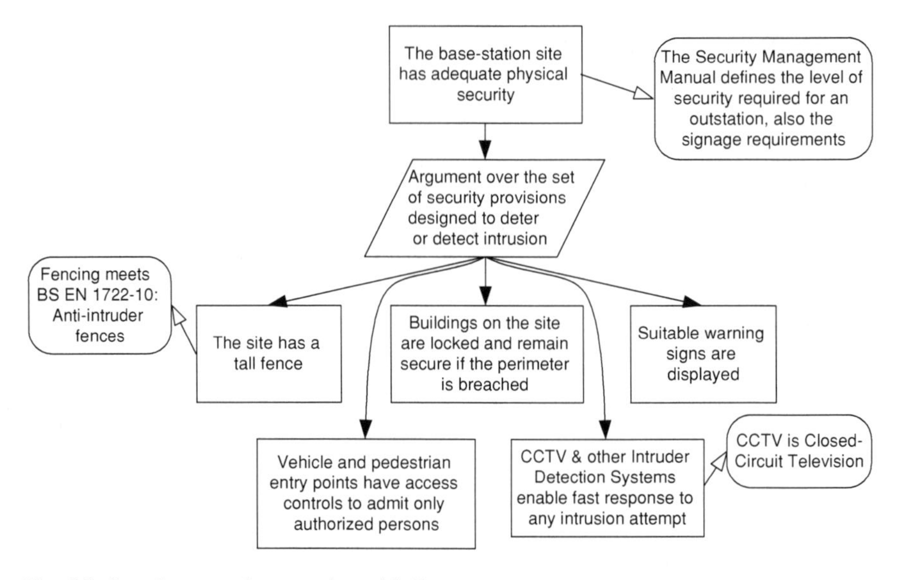

Fig. 6.8 Security example: a version with Strategy

6.6 Using Strategy to Develop the Argument

The process shown in Fig. 6.9, overleaf, provides an approach to developing an argument that requires you to be explicit with your strategy for Goal decomposition. Once you have formulated a good set of Sub-Goals, it gives you the option of suppressing the Strategy, i.e. not using the symbol, if the argument is obvious and emphasis is not called for.

The process is straightforward to apply. The most difficult step is probably the first. The key is to think carefully: What is it that you want to argue; who is it that you want to persuade, and why? Once you have answers to these questions you will be well on the way to expressing a good Top Goal.

Note: This process was not consciously based upon Kelly's Six-Step Method (Kelly 1998) but, on comparison, it is very similar. Following this approach should give more-robust arguments and it forearms you to handle questions from reviewers. You have convinced yourself of the rationale for, and the validity of, the set of Sub-Goals, rather than just bringing them together in an ad hoc manner and hoping for the best.

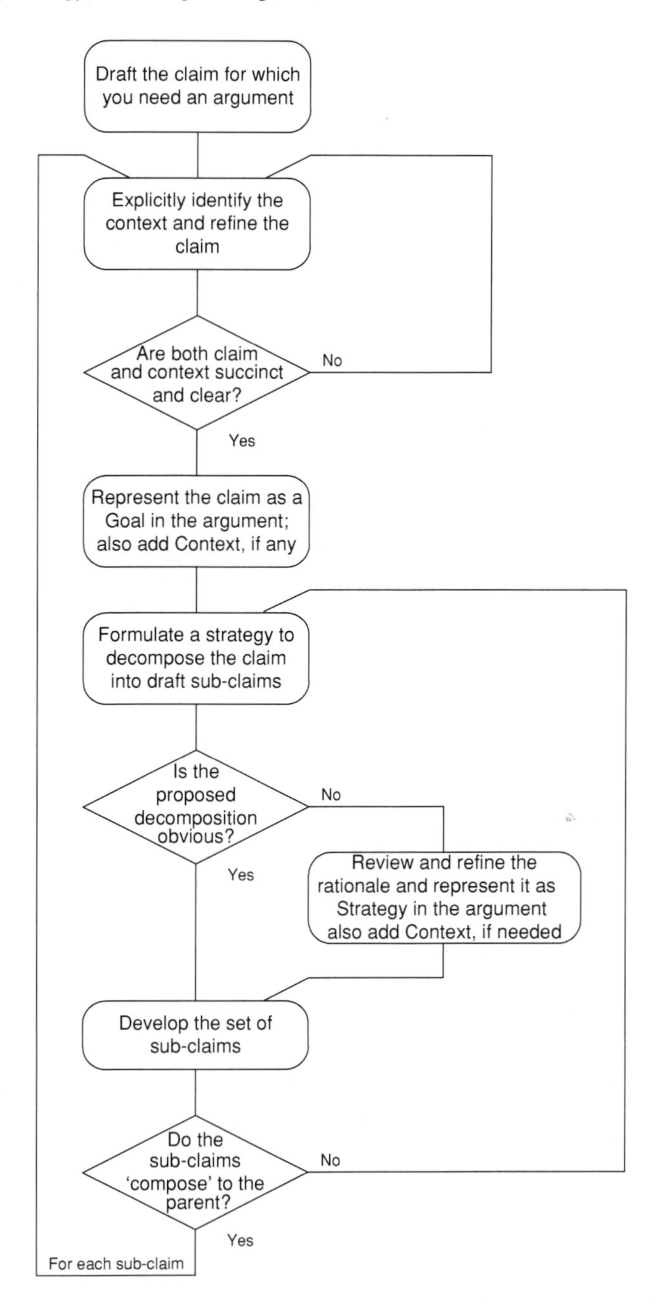

Fig. 6.9 A process for developing an argument

6.7 Reviewing Tip

If you are a reviewer, you can often recognise an ad hoc approach by finding variations on this argument structure: A is true because both B and C are true but, if one of them turns out not be true, then D saves the day. Figure 6.10 shows an example of this.

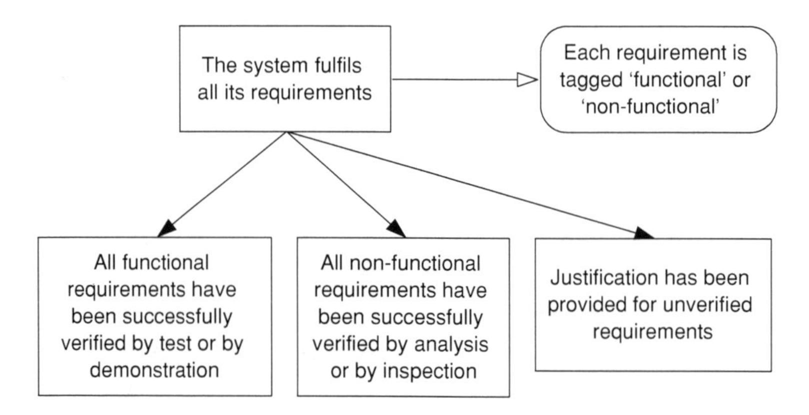

Fig. 6.10 There is something amiss here...

In Fig. 6.10, the left hand and centre Sub-Goals are sufficient to support the Top Goal as all requirements are either functional or non-functional. The right hand Sub-Goal should thus give you cause for concern. I contend that if the decomposition strategy had been made explicit during development, the third Sub-Goal would not be supporting that claim, although it may appear elsewhere in the overall argument. Unverified requirements, and what, if anything, can be done about them, will turn up again in the examples of Chaps. 12 and 13.

6.8 Span of Sub-Goals

The number of Sub-Goals into which a Goal is decomposed is known as its span.

We have looked at simple structures, but complex ones are also possible with GSN. It is easy to conceive of a decomposition of a Goal into a hundred Sub-Goals, say. If we were decomposing the left hand Sub-Goal of Fig. 6.10, for example, we may choose to have a separate Sub-Goal for each functional requirement. But, would we want to? Would it make the argument clear?

In practice we want to reduce the span to a manageable number. Some Human Factors specialists suggest seven plus or minus two (Miller 1956), but even that may be too many in this context. One limiting factor if you are publishing your arguments is the space on a page. A Goal Structure that is of a similar aspect ratio to your page looks better than one that is very wide and shallow.

You can, of course, change the shape of your Goal Structure just by repositioning argument elements, but you can aid clarity by using intermediate Sub-Goals, or multiple Strategies, to reduce the span by adding more hierarchical levels. As this is a Chapter about Strategy, the latter approach is depicted as an example in Fig. 6.11.

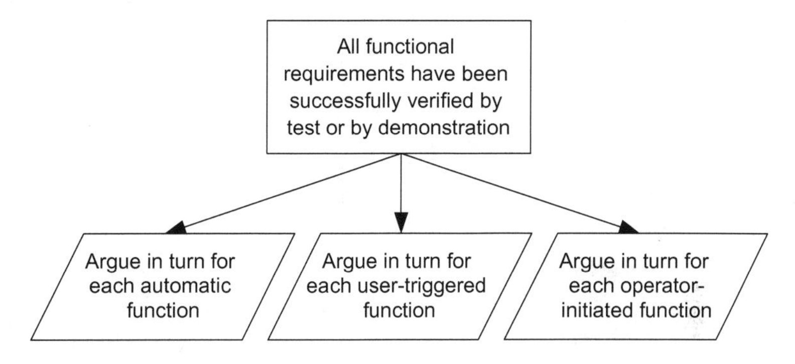

Fig. 6.11 Reducing span by using Strategy to add levels

6.9 Questions

- Try out the process of Fig. 6.9 on a real example. Remember the first step is to ask: What is it that I want to argue; who is it that I want to persuade, and why?

6.10 Problems

1. Look at Fig. 6.11; is there anything missing from the top level of this Goal Structure?
2. How would you express the example of Fig. 6.11 using Sub-Goals, rather than multiple Strategies?

References

Kelly TP (1998) Arguing safety—a systematic approach to managing safety cases (doctoral dissertation). Department of Computer Science, University of York, September 1998

Miller GA (1956) The magical number seven, plus or minus two: some limits on our capacity for processing information. Psychol Rev 63(2):343–355

Chapter 7
A Bit More Contextual Information

Abstract Although the notation introduced so far can be used to represent practical arguments, additional rationale is often required. This chapter introduces the symbol and text convention used to justify elements of the argument. The Justification may also point out, by reference, to an external argument; for example, I may say that a particular Sub-Goal is true because someone has already set that precedent elsewhere and it is applicable in my context. When presenting an argument, it is very important to state explicitly all of the assumptions that you have made. This chapter describes the Assumption symbol and the text conventions for stating assumptions.

7.1 Justification

Justification provides extra explanation or rationale; a reason for what you have done. It is represented by an ellipse annotated with a J at the bottom right. This is contextual information, and so is always associated with the element being justified using an open headed arrow as shown in Fig. 7.1.

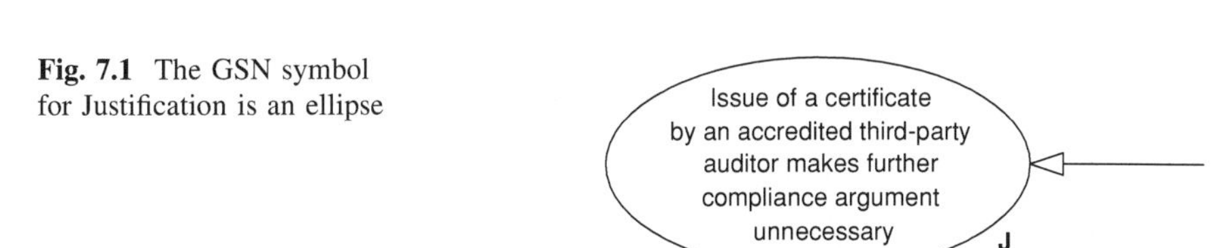
Issue of a certificate by an accredited third-party auditor makes further compliance argument unnecessary
J

Fig. 7.1 The GSN symbol for Justification is an ellipse

Justifications shall only appear at the head-end of arrows, do not hang other contextual information from them. Ideally, for clarity, a Justification should be associated with only one Goal or Strategy, but it may be used to justify more than

J. Spriggs, *GSN—The Goal Structuring Notation*,
DOI: 10.1007/978-1-4471-2312-5_7,

one if required (using a similar configuration to the shared Context shown in the problems to Chap. 5).

The Justification elements in your arguments should be labelled using the same convention as you are using for Context so, for example, if you have decided to label the Context that goes with Goal G1.1 as C1.1, you should label the Justification of that Goal as J1.1.

The text of a Justification is not so tightly constrained as that of a Goal. It may just be a brief exposition explaining, for example, why these particular Sub-Goals have been chosen when others may have been more obvious.

Alternatively, a Justification may contain a simple, "If A then B" argument. It may point out (by reference) to an external argument, "Someone Else's Goal", maybe one that has yet to be presented. It may be a case that is already accepted, "a precedent"; in a legal argument, for example, I may say that a particular Sub-Goal is true because it has already been accepted in the judgment of another case. I would also have to argue that the precedent is applicable in my context.

7.2 Further to Chapter 6

I identified the need for Justifications from the "Option C is the best of the Candidates" example in Chap. 6. I wanted to expand upon the Strategy, to emphasise that, if C is better than all the others, then C is the best. In Fig. 7.2, I have added a Justification (of the "If A then B" form) to the structure; the labelling convention followed in this example dictates that it is to be labelled J0.

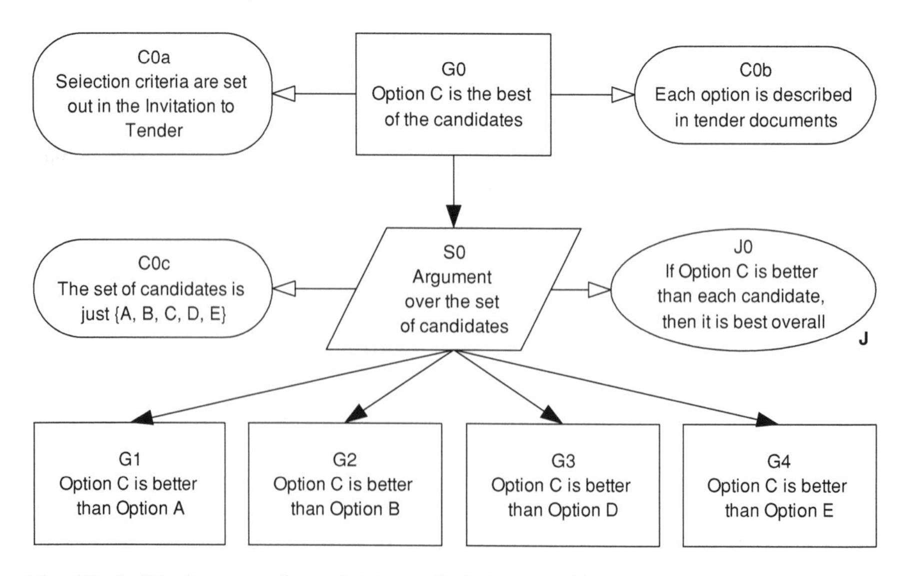

Fig. 7.2 Justification example: explaining a Goal decomposition

7.3 Further to Chapter 4

Chapter 4 spoke of making arguments easier to understand by augmenting the Sub-Goal text with a bit of explanation. That approach goes against the principle of Goal simplification that prompted the introduction of the Context symbol.

Just as a reminder, and for comparison, Fig. 7.3 shows the version with Strategy from Chap. 6.

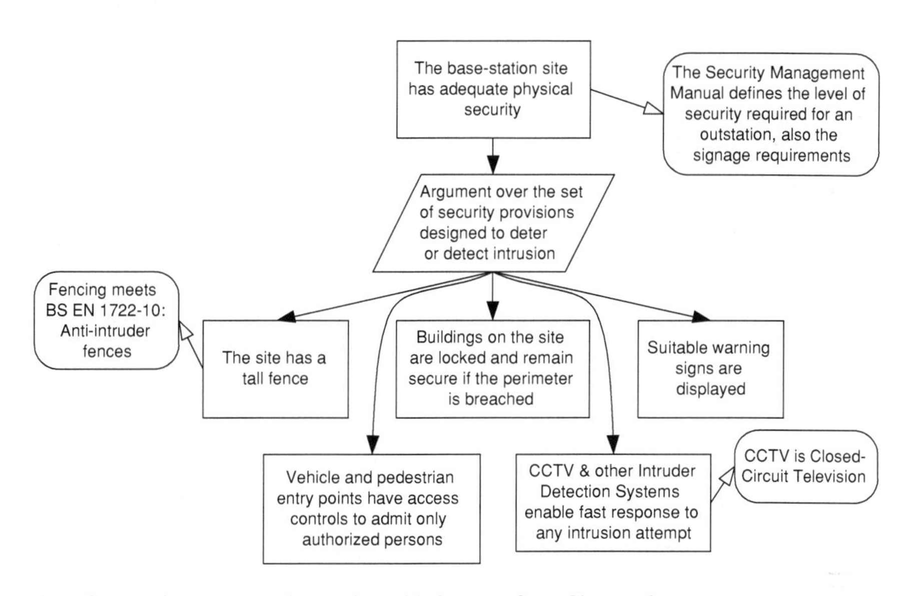

Fig. 7.3 Security example: the version with Strategy from Chapter 6

We can instead use Justification to carry the explanation. I have added Justifications to the version with Strategy from Chapter. To make this new version more legible I have rotated and enlarged it to fit on a page, see Fig. 7.4. Note that I have simplified the text of the Sub-Goals and, consequently, taken the opportunity of expanding the "CCTV" and removing the associated explanatory Context.

7.4 Further to Chapter 2

I also refer to this chapter from the answers to problems posed in Chap. 2. In the answer to Problem 9, I was discussing Goal statements that are valid, but about which there is really no argument. Some such are definitions, rather than claims, and can be represented as Context (See Chap. 3); others are just stating facts, axioms, or self-evident truths. Justification can be used for this; use an assertion,

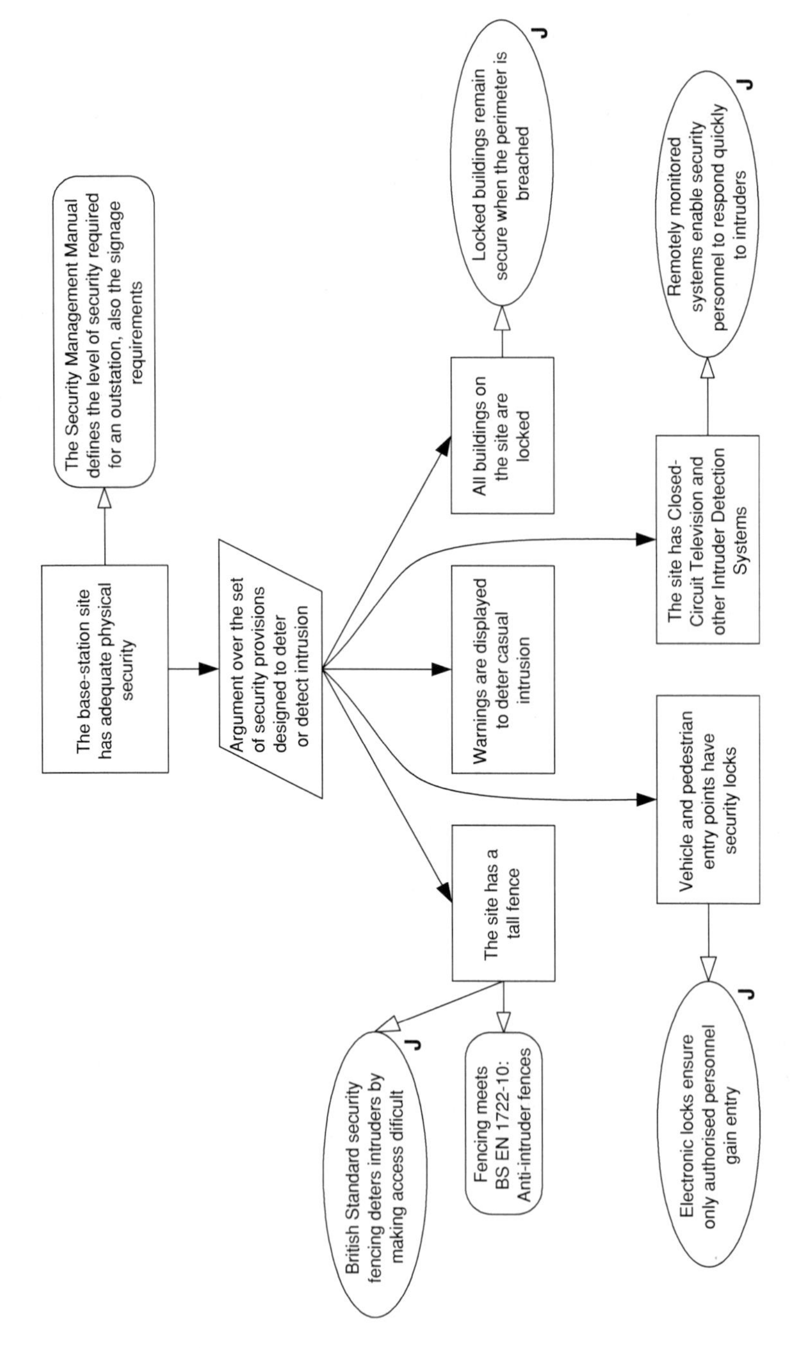

Fig. 7.4 Security example: using Justification (rotated to fit)

rather than the simple argument or explanation usually associated with the Justification symbol.

For example, in an argument demonstrating how effective a fire suppression system is, you may wish to claim, "A fire needs heat, fuel and oxidizing agent". Although you could represent this as a Goal, it will probably be well known to your readers, accepted as a fact, and so a Justification would be more appropriate.

Sometimes you will find yourself wondering whether something requires a Context or a Justification. The decider is why you are adding the information to your argument. Consider the (informal) definition, "A regular icosahedron is a solid with twenty equilateral triangular faces, all the same size". Express this in a Context if an icosahedron has just turned up in your argument (well, you never know) and you want to say what it is. Use a Justification if the icosahedron is already established in your argument, but you want to explain why you have started discussing triangles (or twenty).

In summary, if the definition is used to explain a term, use a Context, but if it is used to explain a decomposition or a strategy, use a Justification.

7.5 Appealing to an External Argument

I could have added to my Business Plan example (see Chap. 5) a claim that properly following the given procedure will produce a correct result. If the procedure authors have done their job properly, an argument should already be available (not necessarily presented in GSN). I can use Justification to refer out to that existing argument, as shown below in Fig. 7.5.

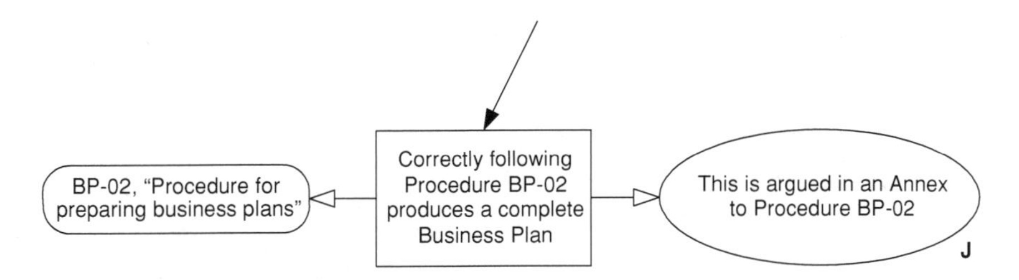

Fig. 7.5 Using Justification to link to an external argument

You can use Justification in a similar manner to partition your argument. In Chap. 6, we had an assurance argument example with two Strategies; one provided an argument for normal operation, the other for failure conditions. The first part would be concentrating on the Concept of Operations, as implemented by the operators, users and their procedures, with some support from the equipment designers and the maintainers. The other part would concentrate on the functional specification as implemented by the equipment and on the maintainers with their procedures, with some support from the operators. If the two parts of the argument are to be produced by different groups, which is a sensible approach given the different emphases, it would be easier to have a document in (at least) two separately published Volumes (Fig. 7.6).

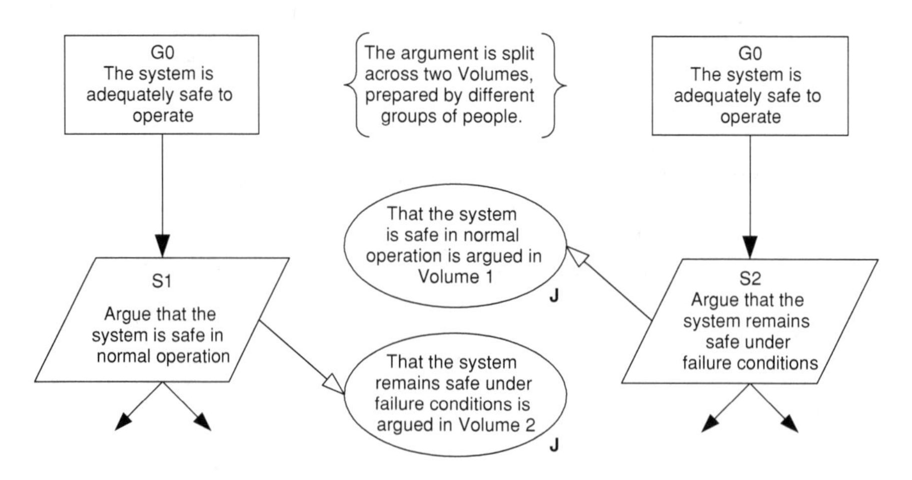

Fig. 7.6 Using Justification to split an argument between Volumes

The link between the two sub-arguments is made with Justification, as shown in the figure. The left-hand segment would appear (with Context as shown in the original figure in Chap. 6) in Volume 1, the right-hand in Volume 2. A similar construction could be used for the Romano-British temple example in Chap. 6; the geophysicists and the artefact specialists could publish their results and arguments in different parts of a journal.

7.6 Simplifying Assumptions

Another type of contextual information is the Assumption, which is something taken to be true for the purpose of the argument. There are (at least) two types of assumption used in arguments.

The first is when you make an assumption to simplify a situation. The argument may be invalid in general, but perfectly adequate in the conditions you assume. That is why simplifying assumptions are regarded as a type of context.

In an engineering environment, I may be arguing that system will be available when it is needed, but this argument may depend upon an assumption that a spare part is held locally and that someone will be there to fit it within an hour of failure.

This example shows that it is very important to state your assumptions explicitly, so that the reader knows you are making them and can judge whether they are reasonable under the circumstances. If my system is deployed in a permanently-staffed location, with a comprehensive stores facility, the argument may well be acceptable. If, instead, the system were to be deployed on top of a mountain, miles from anywhere and with the access road often closed due to adverse weather, an argument with such an Assumption would be rejected.

Assumption is represented by an ellipse annotated with an A at the bottom right. Like the similar Justification symbol, this is contextual information, and so is always associated with the element about which you are stating an assumption using an open headed arrow, see Fig. 7.7. The arrow shows that the Goal, or Strategy, should be read in the context of the Assumption.

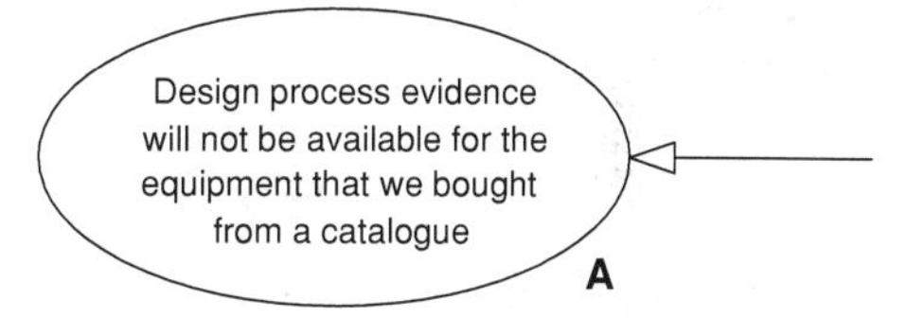

Fig. 7.7 The GSN symbol for Assumption is an ellipse

As Assumptions are contextual information, you should label them using the same convention that you are following for Context and Justification, except that you will use an A instead of a C or J. Assumptions shall only appear at the head-end of arrows, do not hang other contextual information from them. An Assumption can, however, be at the head of several arrows, i.e. providing context for more than one goal, but that makes labelling more difficult…

The symbol is, in effect, saying to the reader, "I assume that…", so there is no need to repeat that in the text. The text in an Assumption should be like that of a Goal, i.e. a succinct statement that can be either true or false; the assumption will be that it is true.

7.7 Assumption Validation

In principle, Assumptions need to be validated to complete the argument, but that validation can be through dialogue with the reader. This may be managed as a two step review in which you ask the reviewers:

- Is each of my assumptions valid (or at least reasonable)?
- Given the assumptions made, is the argument correct?

There is a good chance that you will have your list of Assumptions ready before you have finished the detail of the argument. If you are in a formal Supplier-Customer relationship with your audience, it would improve the chances of acceptance if you were to get them to validate the Assumptions in your list as early as possible; even before the argument is complete.

Delivering a list of Assumptions in advance of the argument puts the onus on you to be clear and context-free in your statements. For example, an Assumption saying, "*Tb* can be taken to be at least 3 in these circumstances" cannot be properly validated without knowing what *Tb* is, and what the circumstances are. If your assumption is that the batteries will keep the system working for at least three hours after a mains failure, say so. Make it explicit and clear. (Also, unless they are absolutely essential, remove *Tb* and all other symbols from your argument.)

Beware of the pitfall of assuming too much. Perhaps your argument depends on the fire alarm being tested at least once a week. Your assumption should be just that the fire alarm is tested at least once a week, and not embroidered with other material. Do not assume, for example, that, "The Ship's Chief Engineer will test the fire alarm every week", when it does not matter to your argument who does it, as long as it is done.

Sometimes assumptions are made at the start of argument development that subsequently turn out to be unnecessary. Go back and remove them. A good way of doing this is to look at each Assumption in turn, and ask yourself, "If this Assumption were to be invalid, i.e. false, would it make the associated claim false?" If the answer is, "No", you do not need to make the assumption; delete it.

Also remove those that have been superseded by facts. At the beginning of an engineering development, for example, you may have had to assume things about the end product, but by the time you need to issue an argument that it is safe to test, say, many decisions will have been made, superseding your assumptions. In some cases you can just delete the fact from your argument; sometimes it can be necessary to retain it, in which case you should make it a Justification, or a Context.

7.8 Assumptions Used in Argument

The second type of assumption you may encounter is used to set up a scenario for an argument of the form, "If I were to assume A is false, it would lead to a contradiction, therefore A is true". Do not use the GSN Assumption symbol for this; the assumption should be stated in a Strategy for the "A is True" Goal (or, alternatively, a Justification). I will go back again to ancient Greece for an example.

Euclid stated a theorem about prime numbers (Euclid 2000) that we can construct an argument for. A prime number is a natural number that has the property that there are only two natural numbers that divide into it without leaving a remainder (natural numbers are the counting numbers 1, 2, 3...). A prime number is thus only exactly divisible by 1 and by the prime itself. Euclid's Theorem is that there are infinitely many prime numbers.

I can construct an argument (Fig. 7.8) assuming that there is a finite number of prime numbers. If this were to be the case, then one of them would be the largest. However, given that largest prime number, I can easily construct from it a larger prime number, giving a contradiction.

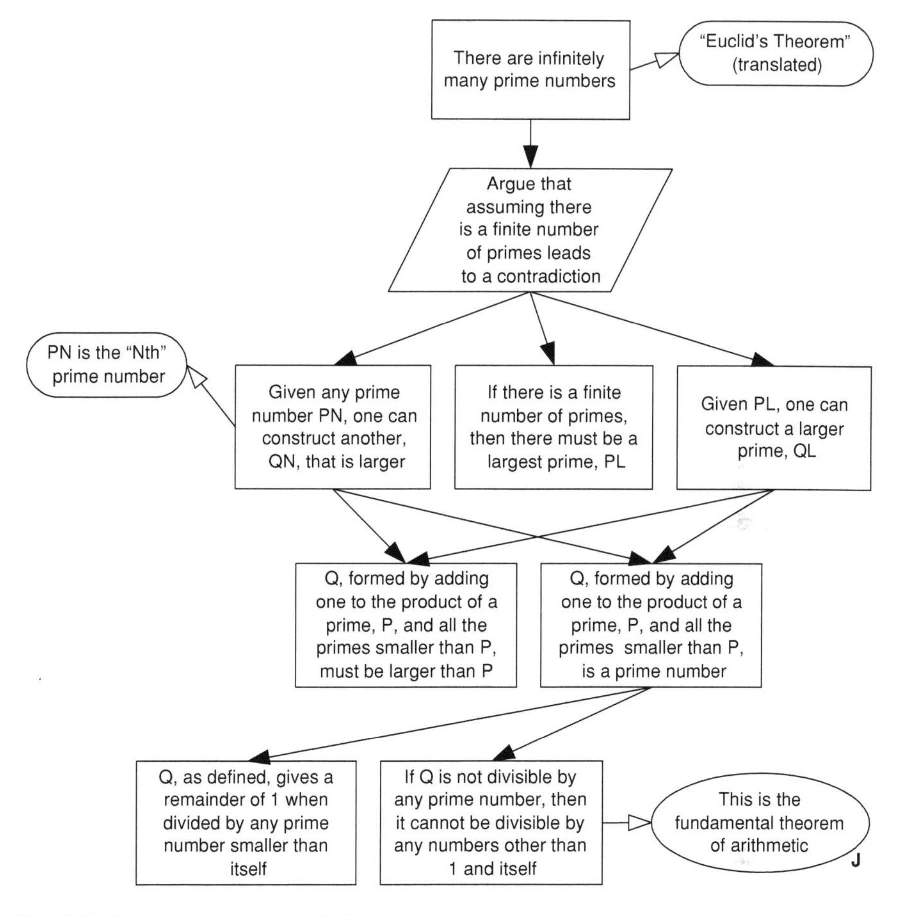

Fig. 7.8 An argument by contradiction

7.9 Questions

- We have a claim, represented by a Top Goal, set in context; we have formulated a strategy for breaking this claim down into simpler components; and we have drawn up the Goal Structure, with Justifications and Assumptions where appropriate. But–something is missing, not all the lowest-level Sub-Goals are self-evident. What is needed to support them? Would Justification fit the bill in all cases? I discuss this in the next chapter (Chap. 8).
- I mentioned Argument by Contradiction to illustrate that the word "assuming" turning up in your strategy does not necessarily lead to an Assumption symbol.

Argument by Contradiction is probably not very useful to the arguments you intend to make, but would you find Contrarian Argument useful? I can illustrate what this is with an example. The overall objective may be to show (in context) that "The system is safe" but you start by constructing an argument that (in the same context) "The system is unsafe". It is then up to the designer, implementer, operator, user, maintainer, et alia, to refute your argument; the resulting discussions (and changes to the system) should result in a more robust system and the basis of a compelling safety argument.

- Are there any useful arguments to be had featuring a regular icosahedron? Imagine the sequence of regular convex polyhedra ordered by the number of faces; these are known as the Platonic Solids. The icosahedron is the fifth in the sequence, with twenty faces; how many faces does the next one in the sequence have? Can you construct an argument to justify your answer?

7.10 Problems

1. Figure 7.9 below shows a fragment of an argument about traffic lights at a road junction with pedestrian crossings; are the three Assumptions appropriate?

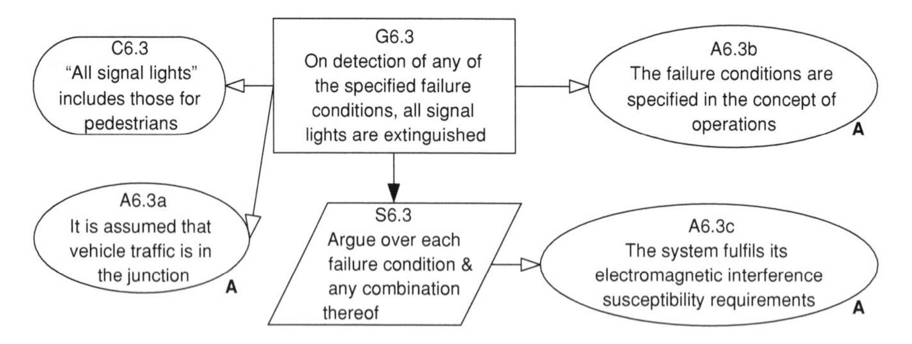

Fig. 7.9 Three Assumptions: do they belong here?

2. The United Kingdom Civil Aviation Authority publishes requirements to be met by organisations wishing to provide services to air traffic. Amongst these requirements is a set of "Sub-objectives" to do with software in systems that support service provision. These are set out in CAP670 SW01 (CAA 2002); we will consider Sub-objective A, "Requirements Validation", as transcribed in Fig. 7.10.

To ensure that arguments and evidence are available which show that the software safety requirements correctly state what is necessary and sufficient to achieve tolerable safety, in the system context.

NOTE 1:These requirements will include requirements to control hazards identified during implementation.

NOTE 2:It is assumed that the system-level safety requirements are derived from a hazard and risk analysis of the ATS environment in which the system is required to operate.

NOTE 3:It is assumed that a necessary and sufficient set of system-level safety requirements exist, which describe the functionality and performance required of the system in order to support a tolerably safe ATS.

NOTE 4:It is assumed that the failure modes which the software must detect and mitigate in order to meet the system safety requirements have been identified e.g. those failure modes associated with: other systems, system-system interactions, equipments, pre-existing software and all user-system interactions.

NOTE 5:It is assumed that the failure modes identified include generic failures relevant to the safety related ATS application, e.g. security threats, loss of communications, and loss of power.

NOTE 6:It is assumed that the failure modes identified (including human errors) are representative of the operational environment for the system and workload on the system operators.

NOTE 7:During the software development process, functions may be introduced which have repercussions on the safety of the ATS system. These will need to be assessed and if necessary, new or changed Safety Requirements will have to be generated.

NOTE 8:The set of software safety requirements includes all software safety requirements derived or changed during the requirements determination and design processes.

Fig. 7.10 Sub-objective a transcribed from CAP670 SW01

So, let us make a start on providing "arguments and evidence"; what is the Top Goal and what context would you put around it? Sketch it in GSN but, rather than capturing all the contextual text, use a shorthand, see the example in Fig. 7.11 below.

Fig. 7.11 Contextual text shorthand for use in this problem

3. Look at the answer of the previous question. Can you see anything wrong with the expanded version, other than the excessive number of contextual information symbols clustered around one Goal? Propose a better arrangement, showing the first level of Sub-Goals.

References

Euclid and Sir Thomas Heath (translator) (2000) The thirteen books of the Elements: Vol 2: Books 3–9, Dover Publications

UK Civil Aviation Authority (2002) Regulatory objectives for software safety assurance in ATS equipment, published as SW01 within CAP670, Air Traffic Services Safety Requirements, 2002 as amended 2010 (available from http://www.caa.co.uk/docs/33/CAP670.PDF)

Chapter 8
The Argument is Incomplete...

Abstract We have seen how to construct an argument in support of a claim by successive decomposition. This cannot go on forever, when have we done enough and how do we terminate our threads of argument? This is the job of the Solution symbol. Solutions constitute the evidence that completes and supports the overall argument. This chapter describes the connection, labelling and text conventions for Solutions, along with methods of justifying how Evidence supports a particular Goal. Sometimes the evidence may actually be another argument; this Chapter identifies a symbol sometimes used to show this, a Goal Developed Elsewhere. It is not part of the core notation, but you may find it in existing arguments.

8.1 The Argument is Incomplete; But What is Missing?

You could argue(!) that an argument is complete when the Goal Structure is completed, it has been accepted as sound, and all the lowest level Goals have been shown to be true. This is a reasonable view. Consider the "Socrates is mortal" example of Chap. 4, which is repeated in Fig. 8.1. The argument is logically correct. It is simple and succinct, but it is not complete as a compelling argument; there is something missing.

I have persuaded you of the truth of the Top Goal, given the truth of the Sub-Goals, but how do I persuade you that the Sub-Goals at the bottom of the structure are true?

Perhaps I could decompose them further, but that is just delaying the problem; I will still have a set of unrequited Sub-Goals.

I can persuade you of the truth of my Sub-Goals by showing you evidence that supports them; it is Evidence that completes an argument. When Goal decomposition has gone far enough, the truth of the lowest-level Goals is demonstrated by appealing to (compelling) evidence. Indeed, "far enough" could be defined on the basis of it being the level at which evidence can reasonably be brought to bear.

J. Spriggs, *GSN—The Goal Structuring Notation*,
DOI: 10.1007/978-1-4471-2312-5_8,

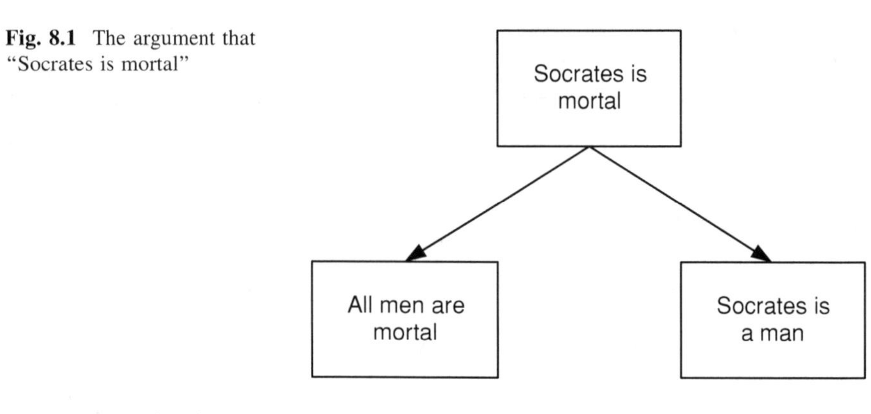

Fig. 8.1 The argument that "Socrates is mortal"

For example, the Business Plan argument of Chap. 5 is complete when the manager has her checklist properly completed. She has physical evidence of completion: there is the plan she was arguing about, reviewed, modified as required, and ready for sign-off; there is also the checklist that constitutes a record of completion, review and acceptance. Usually it is not that easy; the evidence is not all brought together in one place like that. In practice, it would not be unusual to have different pieces of evidence supporting each Sub-Goal.

8.2 Representing Evidence

Evidence is represented in GSN by a circle and it is "appealed to" using a solid headed Thread of Argument arrow, as shown in Fig. 8.2 below. This choice of arrow is to emphasize that Evidence is part of the complete argument structure.

Note: Some authorities call the connection between a Goal and its supporting Evidence the "Is Solved By" relationship, so you may see Evidence referred to as the "Solution of the Goal" in other publications. It is just called "Solution" in many papers; I have called it that in the Abstract of this Chapter and in the caption to the figure below so that those searching for the term will find it. In the rest of the text I will call it "Evidence", because that is what it is.

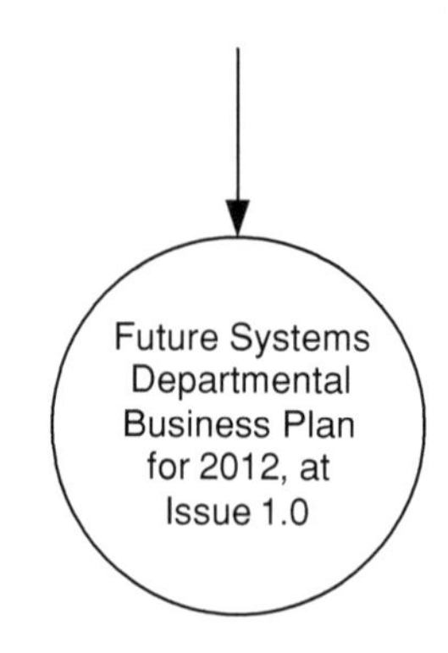

Fig. 8.2 The Solution symbol represents evidence

There will often be more than one item of evidence supporting a single Goal; do not bundle them, use a separate Evidence symbol for each.

I suggest that you label the Evidence in a similar manner to Sub-Goals, but prefixed with an E for Evidence. For example, I would label the Evidence for Goal G1.3.2 as E1.3.2.1 and E1.3.2.2…

It often happens that the same piece of evidence supports more than one Goal; to avoid the confusion that this may cause, some authors label each evidence item uniquely as an alternative to providing cross-references. Using this method of identification results in Evidence symbols not being uniquely numbered in the Goal Structure; this has caused confusion for some readers.

When producing an argument, consider which approach would be clearer for your particular audience and adopt that.

8.3 Specifying Evidence

Note that the Evidence in Fig. 8.2, is not just any Business Plan; it is that for a specific department, for a specific year, and it is a specific version of that plan. If the evidence is a specific instance of something that can change, you need to specify to which version you are appealing. Sometimes the evidence is just a record of something, e.g. a signed declaration; in such a case you need to specify whence to obtain it for inspection. If you intend to return to your arguments in future, it is important to keep copies of evidence items. Otherwise you may find that they are no longer available when you need them.

Sometimes you will want to re-use arguments. I have an argument that a particular product is fit for purpose; the main evidence items are reports from validation and verification activities. If Marketing were to decide that a new version of that product is to be released, I would not want to start my argument again from scratch. It would be better to use the same Goal Structure, but bring new evaluation and test reports as evidence. In that case, I could just have the name of the item in the Evidence symbol, but publish the argument with a table stating exactly which version of which report constitutes the Evidence for this version of the product, and where the records can be found.

Note that it may not be exactly the same argument structure with revised Evidence; the modified product may require a modified argument. For example, I may need to add a Justification of why it is acceptable that the new version has not been fully tested, or that it failed one or more of its tests… These scenarios are addressed in Chaps. 12 and 13, respectively. The former considers what to do if the expected evidence is not forthcoming; the latter suggests strategies for when evidence is found that apparently contradicts your claim.

There may also be the need to modify the argument because a function has been added to, or removed from, this version of the product. It may still be fit for purpose but, presumably, it will not be the same purpose as before. This example illustrates the power of Context; if the purpose for which you claim your product is fit is defined in Context, rather than being distributed throughout the argument structure, it will be much easier to make the required changes.

8.4 But, is it Persuasive?

As I have said before in this book, and will no doubt say again, argument is not proof. The confidence that your readers will have in the truth of the claim represented by the Top Goal depends on the confidence engendered by the Evidence. Sometimes a piece of evidence is sufficient of itself, but often we need to justify it. Why is it appropriate in support of this Goal?

For example, when presenting Evidence in support of the "Socrates is Mortal" argument, you may know that Xanthippe was his wife, but your readers may not, and are likely to question the relevance of her witness statement as evidence in support of the "Socrates is a man" Sub-Goal. If the logical connection between Evidence and the Goal it supports is not obvious, make it explicit with a Justification symbol, as introduced in Chap. 7, see Fig. 8.3.

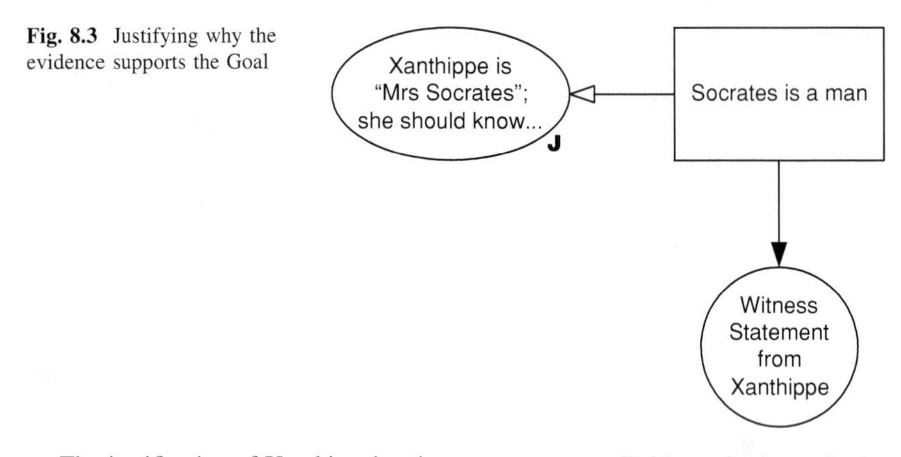

Fig. 8.3 Justifying why the evidence supports the Goal

The justification of Xanthippe's witness statement as Evidence is shown in the figure. Of course, the original claim meant Socrates is one of mankind, but you get the idea. Xanthippe would agree to that; at least she knows that he is not a cat, as concluded by the example of a fallacy in Chap. 4!

8.5 Predictions as Evidence

Our simple example suddenly looks more difficult. The other Sub-Goal is, "All men are mortal". What compelling evidence can you give to support a claim that requires observations over many years to demonstrate? Surely, to be certain of your claim, you will have to wait for all of mankind to die, by which time there would be no point in making the argument, even if you were still around to make it.

In this case I will present a mythical external report that has already done the job of showing that "all man are mortal" for me (Fig. 8.4). As before, it is not sufficient to just wave the report at my readers; I use a Justification to say why it is relevant.

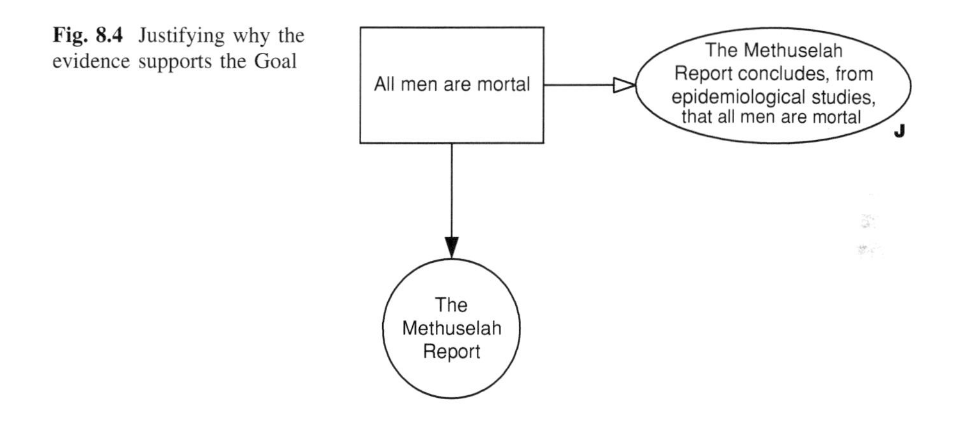

Fig. 8.4 Justifying why the evidence supports the Goal

Figures 8.3 and 8.4 have the same caption, and they look similar, but they show different approaches. Figure 8.3 brings direct Evidence that supports the Goal, whereas Fig. 8.4 is using an external argument as evidence.

Figure 8.4 is not a trivial example; in an Engineering context, a similar scenario arises when I want to show that my new product is reliable. In an extreme case, a systems engineer may want to ensure that her system gives continuous unbroken service. This can be done by providing two pieces of hardware arranged such that, when one fails, the other takes over and maintains the service. For continuity of service to be maintained, the first device needs repair before the second one fails. Therefore, to claim continuity of service, the engineer must argue that the repair can be done before the other device fails (but when will it fail and how long will a repair take?). What evidence can she bring to support her claim?

Similarly, Sales and Marketing want to know when a consumer product will break down. They can then set their 'really generous' guarantee to expire the day before it does so. If they were to derive an average time to failure by observing when all instances of the product cease to function, it would be too late to offer an extended warranty. What evidence can they bring to support the claim that a particular warranty period is the one to adopt?

In both of the above cases, we need to make predictions of when something will fail. Predictions are notoriously tricky (particularly about the future, as they say). If you have prediction-supported goals at the bottom of your argument, readers may agree that your top claim is true, but they may not be confident in it. The Goal is still either True or False, it is just that the reader is less confident as to which. I will discuss justifying evidence more after a brief digression…

8.6 An Alternative Symbol that you May Encounter

In support of the "All men are mortal" Sub-Goal, above, I appealed to a report and justified its use as Evidence by noting that it has our Sub-Goal as its conclusion. In effect, that report contains an argument which has "All men are mortal" as its Top Goal. There is a symbol, available in one of the GSN-supporting software tools, that is sometimes used to show this relationship. I call it the Goal Developed Elsewhere symbol; it is not part of the core notation but you may find it used in existing arguments, see Fig. 8.5.

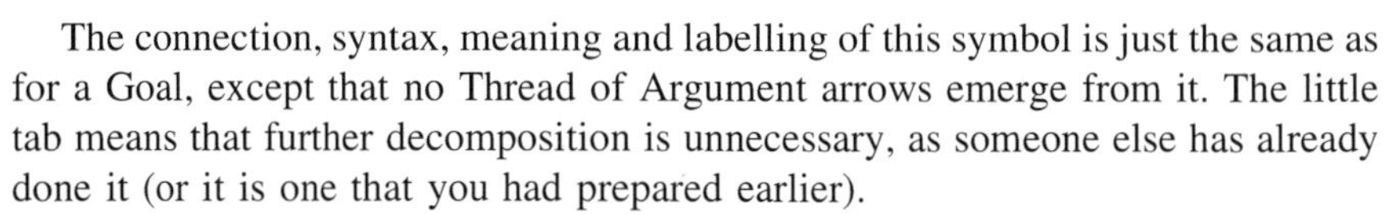

Fig. 8.5 The Goal Developed Elsewhere symbol

The connection, syntax, meaning and labelling of this symbol is just the same as for a Goal, except that no Thread of Argument arrows emerge from it. The little tab means that further decomposition is unnecessary, as someone else has already done it (or it is one that you had prepared earlier).

This symbol is less satisfactory than the Evidence and Justification structure, because it does not tell us where the goal is developed. We therefore need to add some text, stating that the external argument can be found in The Methuselah Report. This could be presented in a table, like that proposed for identifying evidence in "Specifying Evidence" above (or it could use one of the informal symbols that will be introduced in the Note and Labels section of Chap. 9).

Another alternative symbol for using an external argument, the "Away Goal", will be introduced in Chap. 15.

8.7 Avoid the Pitfall of Too Much Justification

Justification of evidence can all begin to snowball if you are not careful. You are ready to present your argument, complete with a bag of Evidence. You do a last review and worry that the readers will have little confidence in your evidence, so you add a Justification to say why they should be confident. You may bring some extra evidence to add weight to this Justification, but is it persuasive enough, should you justify that too? It can get to be like the multiple context example of Chap. 3; you get into a cycle of justification that can erode confidence in the argument, in the same way as a multitude of Contexts eroded the clarity of the argument.

Remember that the Justification symbol often represents an unstated argument. If you end up with too much justification of something, it probably means that you have not taken your argument down low enough. For example, in a continuity of service argument, the system engineer may claim that something is sufficiently reliable, citing a Reliability Report as Evidence, as in Fig. 8.6.

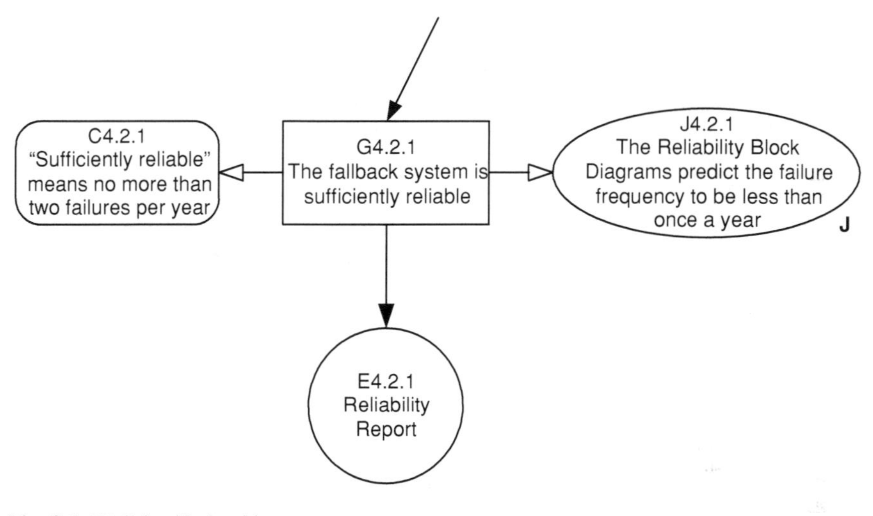

Fig. 8.6 Well-justified evidence

Figure 8.6 is actually an example of good practice using a Justification. The author has not only used Context to define what is meant by "sufficiently reliable", but also stated, in the Justification, the result obtained, so that the reader does not need to read the external report in order to find out whether the target was achieved. That said, there is still scope here for extra justification.

How much extra justification needs to be supplied really depends on the nature of your audience and on the degree of confidence you are trying to instil in them. A manager may accept the figure above as it stands, an engineer may ask whether use of Reliability Block Diagramming is valid in this situation (or vice versa). The Reliability Report should contain that justification, but your audience may require you to declare it explicitly in the argument.

The failure frequency calculations used in the Reliability Block Diagram method depend on the validity of certain assumptions. For example, one of these assumptions says, in effect, that the equipment exhibits random failures, not systematic failures. The failure mechanism is expected to be random component failures, not tolerancing errors in design, for example.

Most modern equipment has considerable software content, and the majority of failures due to software are systematic. I am not saying that you cannot apply statistical methods to software-intensive systems, you can; I am highlighting that some underlying assumptions may not be valid in your particular case.

In practice, most reliability engineers would use a software tool to construct, present and analyse Reliability Block Diagrams. It is the tool that calculates the result, so is it trustworthy even if the underlying assumptions are valid? We need to justify that this particular tool is appropriate for use in this case. Then we may question whether the tool was used correctly, and feel we need to justify that too.

If you were to add all that to the diagram, there would now be four Justification symbols clustered around the Goal. Rather than being reassured by this, the reader may begin to think that you are unsure of your argument (when, in reality, you were unsure of their readiness to accept your argument). The problem is that, for this audience, we have not argued in sufficient detail. We need sub-claims, one of which will be the Justification from the previous figure.

The augmented argument is shown in Fig. 8.7. Note that I have not used a Strategy to explain the argument, because I believe that it should be clear to the reader. I argue that a particular analysis method is valid; the tool used to implement this method is fit for purpose; those using the tool are competent so to do; and the result they produced thereby exceeds the target by a factor of two.

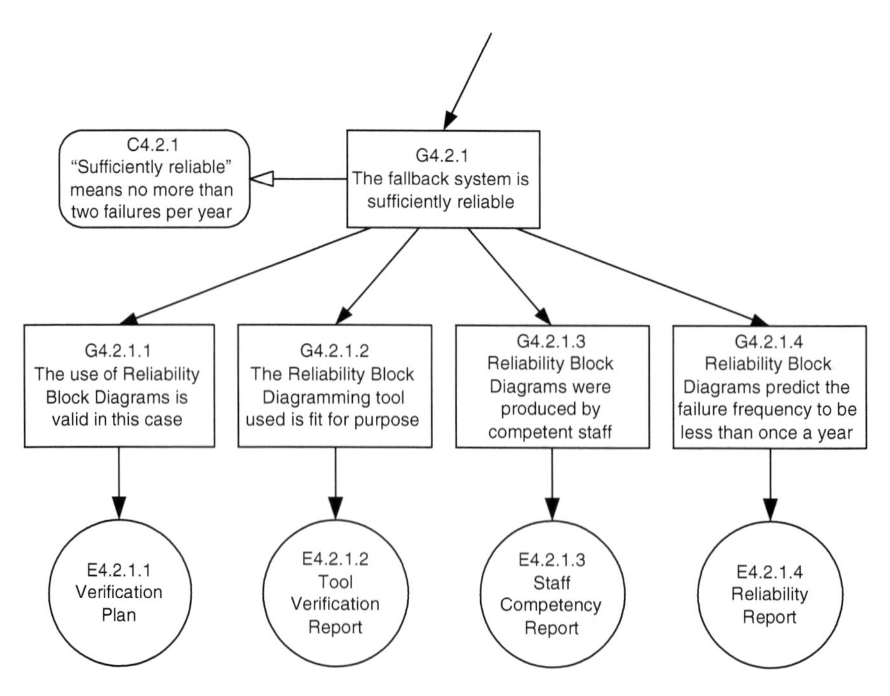

Fig. 8.7 Well-supported Goal

It may be more compelling, for some audiences, if the declaration of the result came first and was positioned directly below the target stated in the Context. But get the basic argument down first; rearrangement and clarification can come later. It is important not to get bogged down in detail in one thread of the Goal Structure before drafting the higher levels of the other threads.

Note that Chap. 14 looks at argument structures for reasoning about processes being fit for purpose and implemented competently; it also considers how such arguments could be applied to tools.

8.8 Avoid the Pitfall of Incomplete Evidence

We have looked at over-justifying evidence; the reverse can also happen. The evidence looks really good to you, so you do not justify it, but in reality it does not properly support your claim.

A silly example is Xanthippe's assertion that Socrates is not a cat. It supports the Goal that Socrates is a man, but it is not sufficient. A strategy of eliminating every other genus is impractical; positive evidence would be much more compelling. Other situations may be less obvious; you may have found an impressive amount of evidence demonstrating that something worked on previous occasions, but in many cases that will say little, or nothing, about future outcomes. This was well illustrated by Richard Feynman in an appendix of the Rogers Commission Report on the loss of the Space Shuttle Challenger (Rogers 1986); he wrote: "When playing Russian roulette the fact that the first shot got off safely is little comfort for the next".

This situation comes up frequently in engineering when we want to argue that a piece of equipment will do the job at hand, based on our experience of using this equipment before; a "previous use" argument. If we want to reuse something that we bought from a catalogue, we will have little or no information upon which to build an assurance argument, but we do have evidence from our prior usage. Check—is the evidence pertinent; are you comparing like with like and building a compelling case, or are you comparing apples and pears and getting a lemon?

For example, is the item you can buy now really the same as the one you bought before? Many suppliers 'improve' their product but sell it as if it were the same item. If it is the same thing, is the environment in which you will use it the same as last time? By environment, I do not only mean physical conditions like the ambient temperature and vibration, but also such things as operator usage profiles, mix of input signal types, numbers of network connections, etc. Will you use it for the same purpose, and in the same way, as before? The computer that you found perfect for reading documents may not be suitable for viewing movies.

For a compelling "previous use" argument, even if the equipment and usage environments can be argued to be sufficiently similar, you will still need to show that your data are complete and comprehensive (Fig. 8.8). You would need, for example, to present an analysis of all of the incidents experienced in the previous use, not just some of them. You therefore need to argue that the reporting process was good enough to record all pertinent incidents. The incident reports need to distinguish between equipment failure, or malfunction, and other causes. For example, does "Loss of service due to power supply failure" mean that there was a power cut, or that power supply components within the equipment failed?

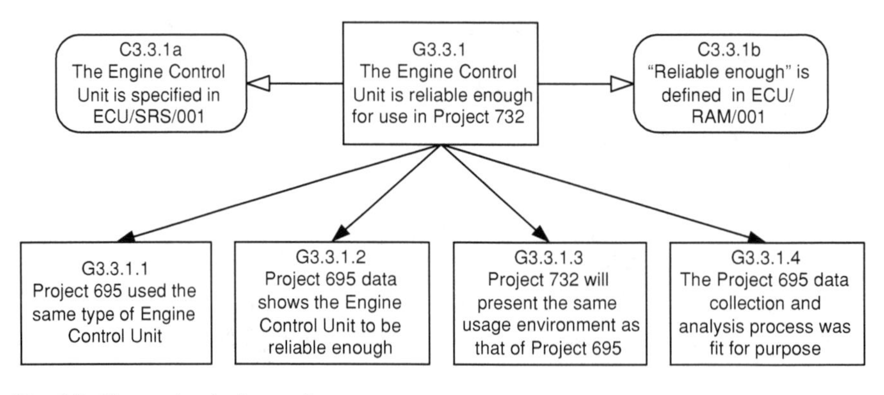

Fig. 8.8 The top level of a previous use argument

If you are arguing any more than why to buy this item, rather than some untried alternative, a "previous use" argument is rarely compelling on its own. Normally, you will also need to bring additional assurance, an argument over comprehensive testing in the new context, for example.

8.9 Avoid the Pitfall of Inappropriate Evidence

Justifying that your Evidence is sufficient can be difficult, as can showing that it is correct and pertinent, but you can succeed. Your readers may admit that the set of evidence you presented does support your claim, but they may then question its provenance.

For example, a car salesman is presenting an argument that his product is just the vehicle that his Customer needs for her fleet; the Customer will be wondering whether the performance report he offered as evidence came from a production example, or from a special version specially tailored for fuel efficiency. The salesman needs to present evidence supporting the evidence. In this case it may be a formal report from an accredited "test house", for example.

The pitfall is that, in general, you cannot obtain these additional evidence items retrospectively. You need to at least draft-out your argument in advance, so that you can plan-in the production and collection of evidence.

Where evidence has been generated or obtained by a defined process, as should be the case in scientific, engineering and legal applications, a separate argument can be made about the process. This may be an external argument linked via a Justification to your main argument, as in Fig. 8.4, or it may be a distinct "leg" of your argument structure, as in Fig. 8.8.

In a process argument, it is usual to argue that there is a process that is adequate for the task at hand and that the process has been properly followed in this case. To be amenable to such argument, a process should be clearly documented and generate auditable records of some kind.

It may also be necessary to argue for the competence of the people who followed the process and for the suitability of any tools they used in doing so. See Chap. 14 for examples of arguing that tools and processes are fit for purpose and used competently.

Formal change control processes are used in most engineering environments that require arguments to be produced. Records from these can support the claim that, for example, the analyses reported upon were indeed done on the version of the design that is the subject of the argument.

If your evidence does not arise from use of a documented process, you should try to identify other means of recording its provenance. Photographs could be used for this, for example, if the car salesman did not have a report from an accredited source, he could have presented photographs of the tests in progress. Other ways of showing provenance include signed declarations, or certificates; independent "second opinions"; and widely-published, peer-reviewed, sources.

8.10 Questions

- We have looked at examples where Justification is used to show that the offered evidence adequately supports the claim. What about situations where this is not sufficiently compelling? For example, imagine a scientist makes a claim based on measurements obtained from an experiment: "These measurements show the X Effect, as predicted by Y Theory". Other scientists read her paper and agree that these results are consistent with that effect, but claim that they are more likely to have arisen from other mechanisms not related to Y Theory. Justifying the evidence further would not persuade these detractors; what else could the scientist do in support of her claim of having observed the effects predicted by Y Theory? A similar scenario is discussed in Chap. 10; it is the first lesson to be drawn from an extended example.
- Another potential pitfall is that, sometimes, you get more evidence than you bargained for. "Counter-Evidence" is evidence that does not support your claim; rather it is evidence that contradicts it. What, if anything, could be done to save the argument in these circumstances? I shall return to this topic in Chap. 13.

8.11 Problems

1. In the examples, I use E for Evidence to prefix the numbers of Evidence nodes; can you see where confusion could arise if we were instead to use the prefix S for Solution?
2. You are updating an existing argument and have been asked to ensure that it uses only "core GSN" symbols. You find deprecated symbols representing Sub-Goal G2.3.1, see Fig. 8.9. How should this be redrawn?

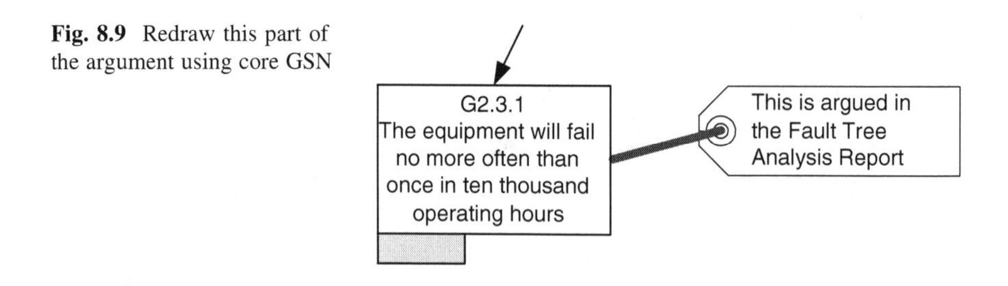
G2.3.1
The equipment will fail no more often than once in ten thousand operating hours
This is argued in the Fault Tree Analysis Report

Fig. 8.9 Redraw this part of the argument using core GSN

3. If you did the last problem, you probably had the Fault Tree Analysis Report as Evidence, supporting the Goal. I would hope to see a bit more information in a real argument. Can you see how it could be made more compelling? How does the report show that the target has been achieved, for example?

Reference

The Rogers Commission (1986) Report of the presidential commission on the space shuttle challenger accident. Appendix F—personal observations on reliability of shuttle by R. P. Feynman

Chapter 9
The Argument is Ready for Review...

Abstract Early on in its development, you ought to get your argument reviewed by someone else; preferably your Customer, if you have one. This chapter introduces additional notation that you can use to let your reviewers know where in your Goal Structure there is more to be done. These additions include the Goal to be Developed symbol, which is also known as Undeveloped Goal. Some people like to add explanatory notes to their Goal Structures; this chapter also suggests some symbols for this, but note that they are not part of the formal notation and they can clutter your diagrams. Finally, I start development of a checklist that may be used by the reviewer of arguments presented in GSN; the author of such arguments should always use it...

9.1 How do I Show that the Argument is Unfinished?

If your argument is large or complex, I recommend that you get it reviewed by someone else early on in its development. If you were to wait until it is nearly complete, you could find that there is a lot of rework needed due to a trivial mistake early on. In other words, if you do not develop the argument incrementally, you may find that you have done it excrementally.

To scope the review, you need to let your reviewers know where in your Goal Structure you feel you have done enough, and where there is more to be done. The Evidence symbol (see Chap. 8) shows that you have finished with those threads of argument that terminate on Evidence. We can use the "Goal to be Developed" symbol to highlight unfinished threads. Note that some authorities call this the Undeveloped Goal symbol.

The form of this GSN symbol may be familiar to those proficient in Fault Tree Analysis, wherein the analyst denotes an incomplete branch by a diamond. In GSN, we attach the diamond directly to the Goal, as in Fig. 9.1, which shows its use on a Sub-Goal of the "best candidate" argument from Chap. 6.

J. Spriggs, *GSN—The Goal Structuring Notation*,
DOI: 10.1007/978-1-4471-2312-5_9, © Springer-Verlag London Limited 2012

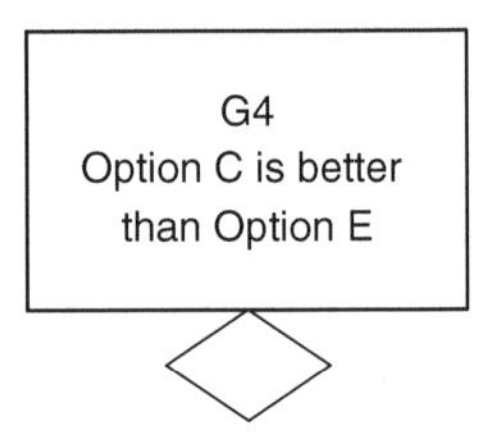

Fig. 9.1 The Goal to be Developed symbol

The syntax, meaning and labelling of the Goal is just the same as before. The addition of the diamond just means that further decomposition has yet to be done. Such a Goal thus only has incoming Thread of Argument arrows (see Chap. 4); the diamond indicates that none come out at this stage of development. You may attach Context and Assumption symbols to an undeveloped Goal, but it would be unusual to use a Justification; there is no decomposition to justify.

You do not have to stop decomposition at a Goal; you may wish to indicate what will happen next by presenting a "Strategy to be Developed", as shown in Fig. 9.2 below.

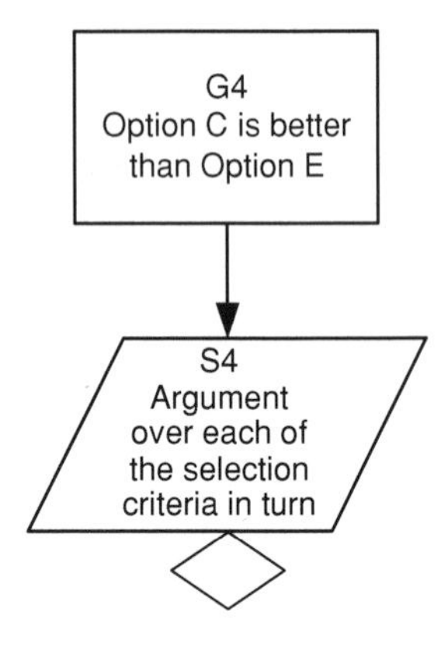

Fig. 9.2 The Strategy to be Developed symbol

In fact, this may be the better option in most cases, as you can get feedback on your intended next steps, as well as on what you have done. Using this symbol in a review copy does not commit you to having a Strategy there in the final version. In a simple case, you may replace it with a couple of obvious Sub-Goals; see Chap. 6 for a discussion of when to use Strategy symbols.

Note that, in a real argument, I would also have used a Context to say where the selection criteria are documented (unless, of course, that had already been shown at a higher level, and so is inherited here).

9.2 Notes

Some people like to add explanatory notes to their Goal Structures. (I slipped one into Fig. 7.6 of Chap. 7; did you spot it?) In a large structure that goes over many pages, you may wish to add a note saying, for example, "This part of the argument addresses the bespoke equipment; bought-in equipment is addressed in Goal 4.3.8". These note symbols are for putting this information on the diagrams, rather than using text between them.

Notes can be presented as bracketed text, or you could use a word balloon or even a "tie-on label" (Fig. 9.3). None of these symbols are part of GSN but, if you find them useful, be consistent and use the same symbol throughout your argument.

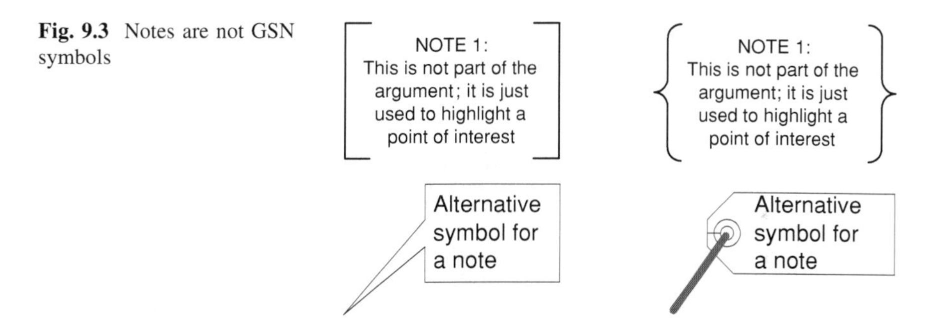

Fig. 9.3 Notes are not GSN symbols

Notes may stand alone on the page, but if you have to associate them with a particular Goal, and you have standardised on one of the bracket forms, use the open-headed contextual information arrow. Be careful when adding a Note to a Goal Structure. Consider: is it adding useful information to aid navigation, or is it highlighting that your argument is less than compelling, or has other flaws?

9.3 Labels for Navigation

If your Goal Structure is split over many pages, you may also wish to label it saying where in the document a particular Goal is expanded. You can use one of the Note symbols, above, but it looks neater to draw an arrowhead on the diagram to say where to look next (Fig. 9.4).

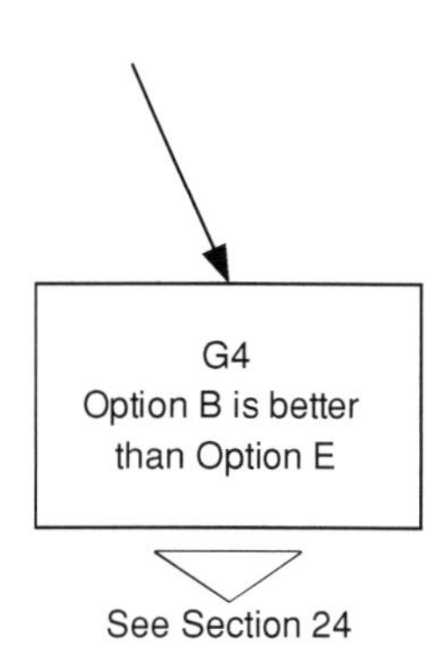

Fig. 9.4 Label showing where to go next—not a GSN symbol

If you do this, it is also a good idea to show whence a Goal came. In Section 24 you could have the following (Fig. 9.5).

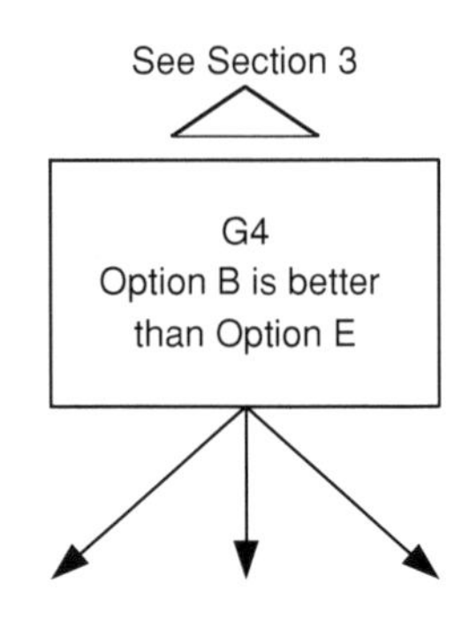

Fig. 9.5 Label showing whence you came—not a GSN symbol

It is not a good idea to use specific page references, unless you can get them to update automatically when the document is changed. A better solution is to give each part of the Goal Structure a section heading from which a Table of Contents can be automatically generated, as was done in the examples above. Alternatively you could use Captions and automatically generate a Table of Figures.

A better option to the use of labels may be to have a table below the diagram with cross-references to where the lowest Sub-Goals are addressed. In most word processors, it is possible to arrange for the cross-references in such a table to be maintained automatically as parts of the document are moved around and others are added or deleted.

9.4 Pause for Reflection

We have now met all the most important symbols in the Goal Structuring Notation. It is time to pause and reflect. What would a reviewer make of our arguments? How can we make them more reviewer-friendly?

I have been suggesting things to make your arguments more acceptable to the reader as we have been going along: succinctly state your Goals; clearly define your terms in Context; use Strategy and Justification to give your rationale; explicitly declare your Assumptions; make full use of Justifications to explain why your Evidence is good; and so on.

I have also given some explicit hints to reviewers. For example, watch out for an "Argue over N cases" scenario where there are N + 1 Sub-Goals, the last of which suggests all is not well with the N (see Chap. 6, Fig. 6.10). We need more; is there some generic checklist for use when reviewing arguments expressed in the Goal Structuring Notation? No; so let's make one (Fig. 9.6). Authors should also apply such a checklist before sending their arguments out for review.

9.5 Constructing a Checklist for Authors and Reviewers

I am aware of a paper on the subject (Kelly 2007) but, before reading it, I will apply the approach I advocated in Chap. 5 and generate a checklist of my own. I can then compare it with the paper and see if I missed anything. In Chap. 5, we had the claim, "My Business Plan is ready for review by the Board". This case is similar, but we do not need to specify by whom our arguments will be reviewed.

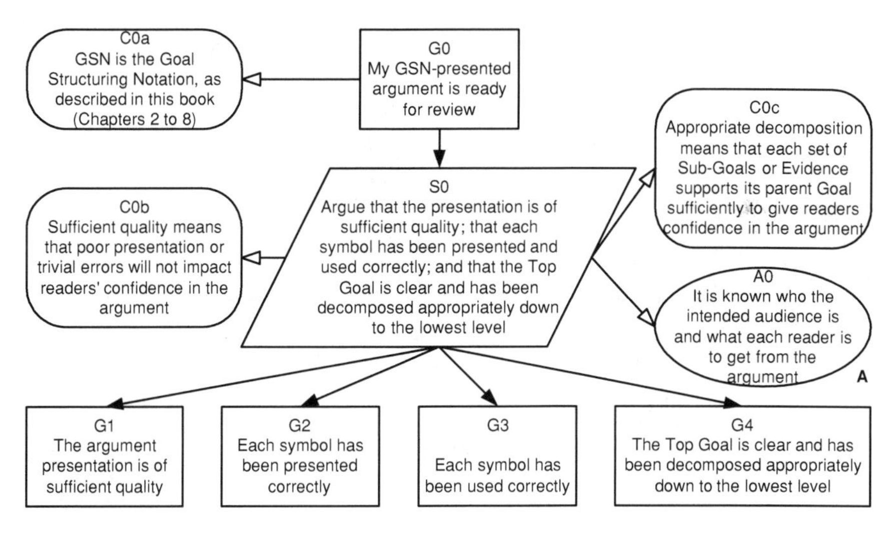

Fig. 9.6 Top level of review argument

9.6 Quality

I hope the rationale for having Goal G1 in Fig. 9.6 is obvious; if you are trying to persuade someone of something, you do not want to distract them with unreadable diagrams, ungrammatical text and lots of typographical errors.

Many drawing tools have a built-in spelling checker—if you have one, use it. Also, use the spelling and grammar checkers in your word processor, first making sure that they are set up properly for your chosen language and grammatical style.

Note that, in Fig. 9.7, I have assumed that the argument is to be issued as a document and that, if you have special templates, document management systems, and such like, you will add claims about those aspects yourself.

Some people seem to 'confuse the map with the territory'; instead of having a document section entitled "The Argument", they present "The GSN" as if it were some infallible oracle. It may be better to use your proposition as the section title anyway; another reason to be succinct…

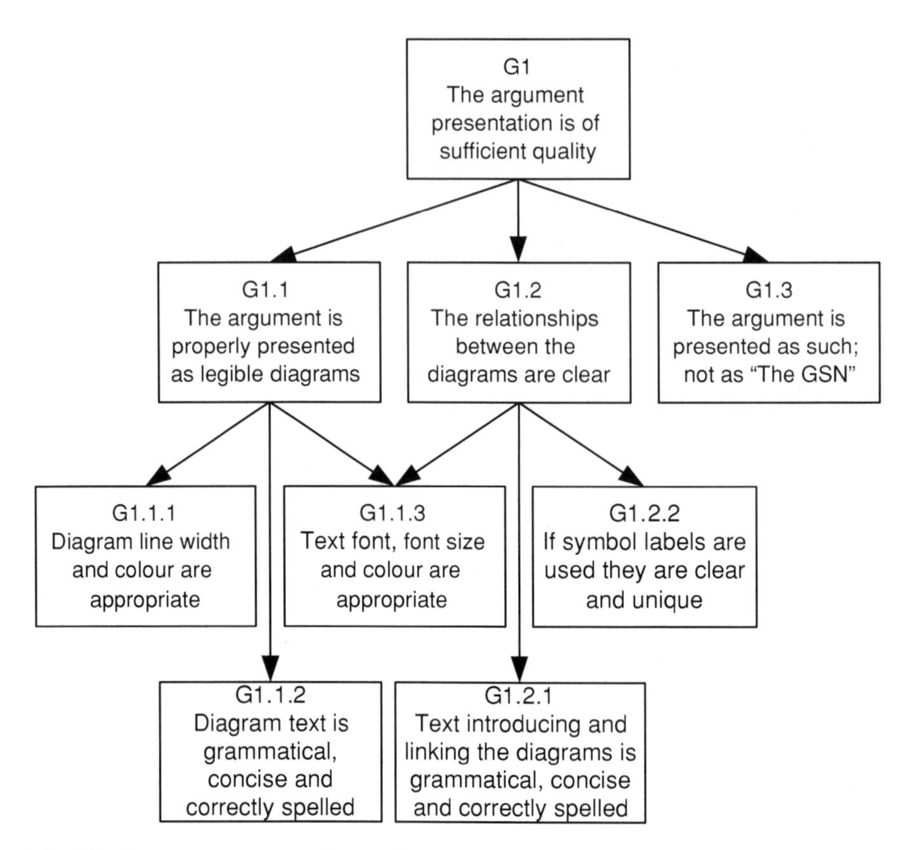

Fig. 9.7 The argument presentation is of sufficient quality

I have been purposefully vague here. What constitutes 'sufficient quality' depends on what you are arguing and to whom. When trying to persuade someone of something, it is better to err on the side of too much quality, rather than too little...

9.7 Correct Symbols

Goal G2 is concerned with correct GSN symbols (Fig. 9.8).

This may seem unnecessary, but I have received examples of arguments for review that had ambiguous symbols. For example one had ellipses without A or J annotation. In some places it was not clear whether the author was justifying something with a confident assertion, or assuming that it was so. It does not give confidence in an argument if you have to start assuming which statements constitute assumptions.

Another review example used hexagonal symbols without any explanation of what they represented. A reviewer should be assessing the argument, not searching for it in your diagrams.

Many of the arguments presented in GSN are assurance arguments. For example, in safety engineering, you may argue to assure the reader that all the necessary steps have been taken to make some machine safe to use. If you start by demonstrating that you cannot even use your chosen notation properly, what confidence will the reader have that you have done the engineering properly?

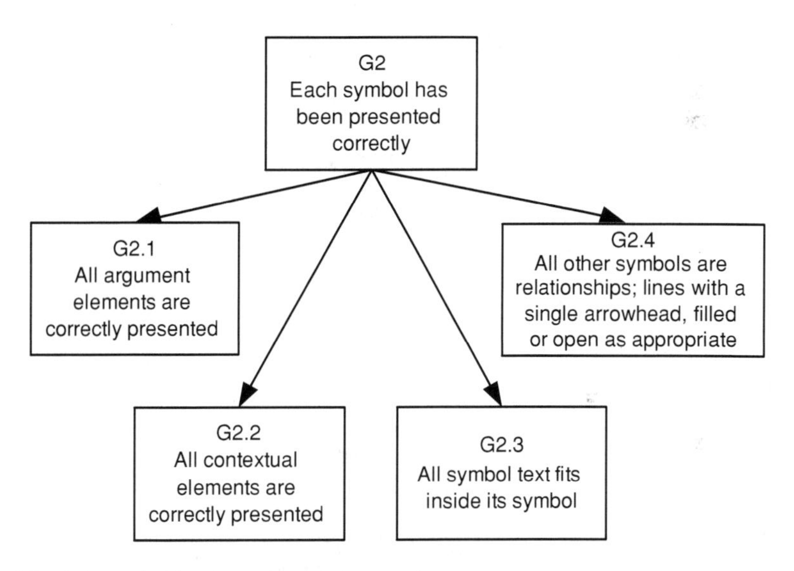

Fig. 9.8 Each symbol has been presented correctly

Sub-Goals G2.3 and G2.4 do not need breaking down further, but the other two can be decomposed to identify each type of element separately. The argument elements are Goal, Strategy and Evidence, whereas the contextual elements are Context, Justification and Assumption. See Fig. 9.9 for the decomposition of Goal G2.1 and Fig. 9.10 for that of Goal G2.2.

Note: There are other symbols for argument elements, contextual elements and relationships in extensions to GSN (See Chaps. 11 and 15). If you use these extra symbols, add pertinent Sub-Goals to this argument to extend your checklist.

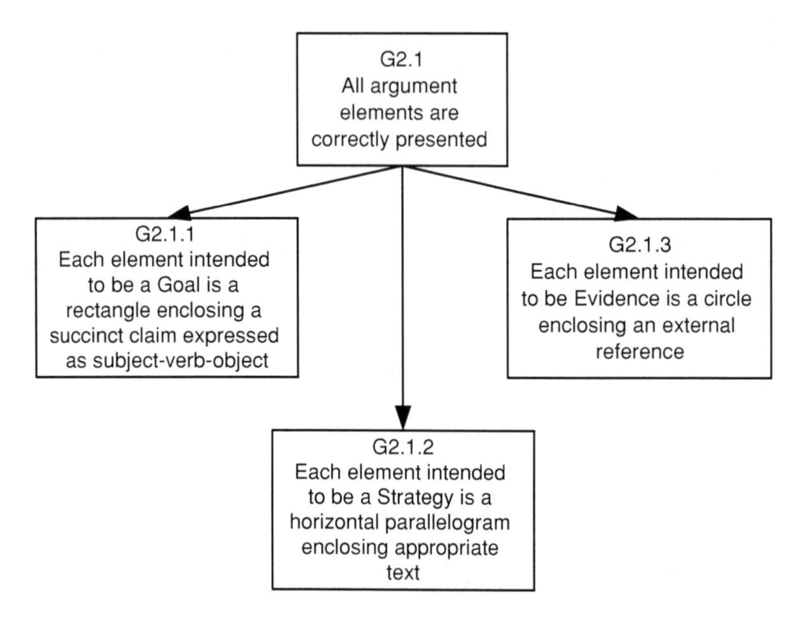

Fig. 9.9 All argument elements are correctly presented

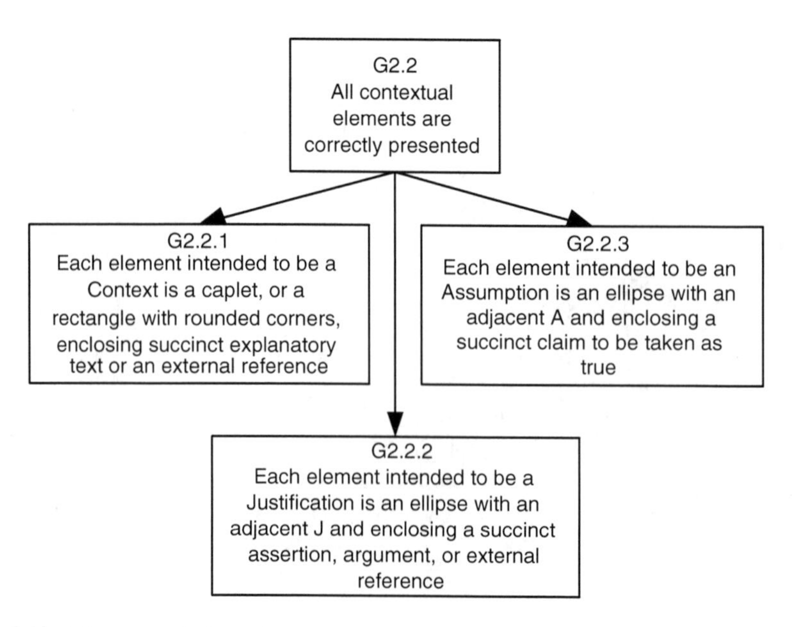

Fig. 9.10 All contextual elements are correctly presented

9.8 Correct Relationships

Now we are sure that we have correct symbols, do they appear in correct relationships with each other?

Is the Goal Structure connected, i.e. are there any bits that are not joined on? This can arise when you modify a Goal's decomposition, but forget to remove any now unwanted Sub-Goal decompositions from elsewhere in the document. Goal G3, shown in Fig. 9.11, checks that all the argument elements and contextual information symbols are shown in the correct relationships.

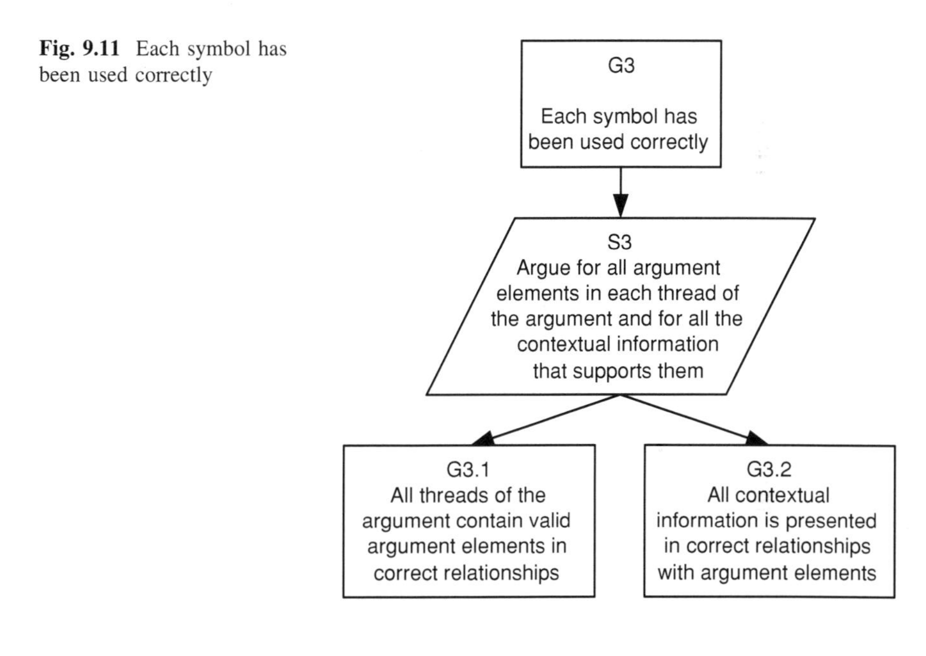

Fig. 9.11 Each symbol has been used correctly

To decompose Goal G3.1, think of the threads of argument; ultimately they all depend from the Top Goal, but what about the other ends, how are they terminated? Each thread runs from Goal to Sub-Goal, maybe via a Strategy, until it arrives at a Goal supported by Evidence, or an explicit termination at a Goal to be Developed symbol. See Fig. 9.12 for my decomposition of this Goal.

Contextual information is different, although its influence flows down the Goal Structure, the actual symbols are all terminators; there are no threads of context.

Clarity is very important for contextual information. Sometimes it is better to have it repeated at different points in the Goal Structure, even though that may make maintenance of the argument more difficult. We need claims that repeated context cannot be consolidated and that shared context would not be better repeated, see Fig. 9.13.

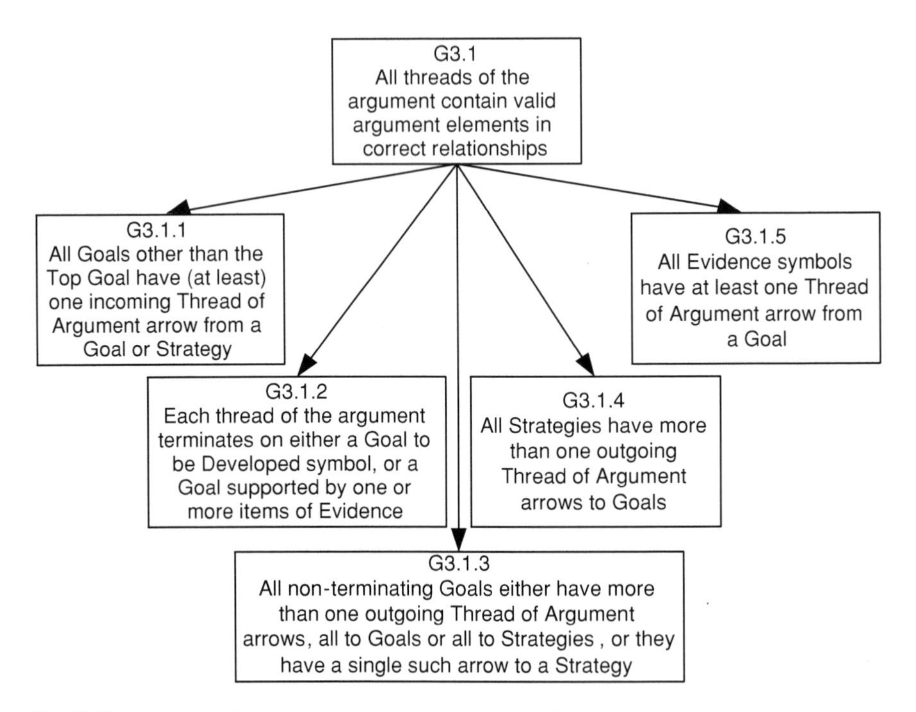

Fig. 9.12 Argument elements are correctly interconnected

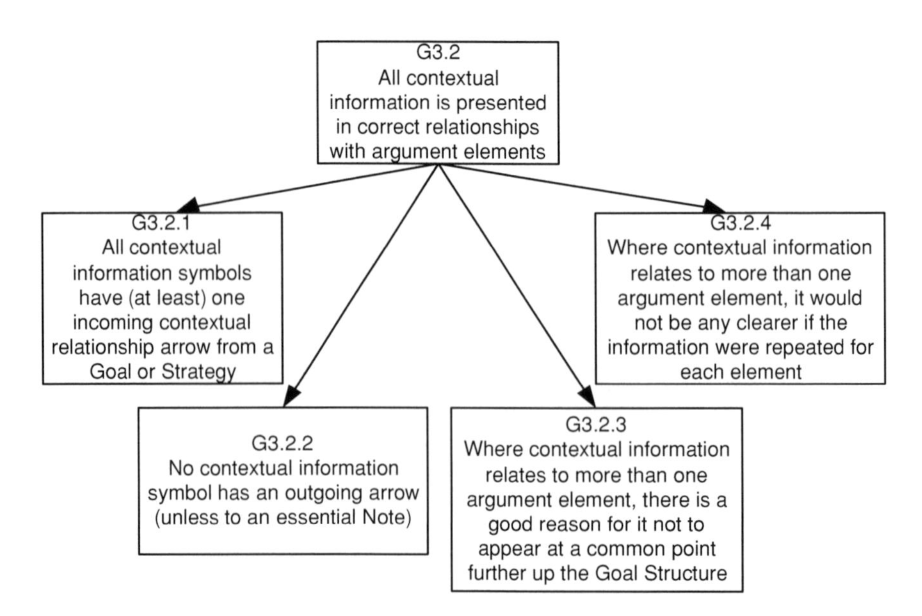

Fig. 9.13 Contextual relationships are correct

9.9 Correct Argument

Goal G4 claims that "The Top Goal is clear and has been decomposed appropriately down to the lowest level" (Fig. 9.14). That Top Goal will have been checked for syntax (and brevity) in support of Goal G2 and the structure of its decomposition down to the lowest level will have been checked in support of Goal G3. What we are looking at here is the meaning of the Top Goal in its context (Fig. 9.15) and how logical each stage of decomposition is (Fig. 9.16); this is both a semantic and a completeness check. In practical terms it means looking at each Goal in turn and checking the decomposition is appropriate and complete.

This is the only Goal that a reviewer should be interested in, because the author will have already fulfilled the prior Goals of Quality, Correctness of Symbol and Correctness of Interconnection. Of course, authors should use this one too...

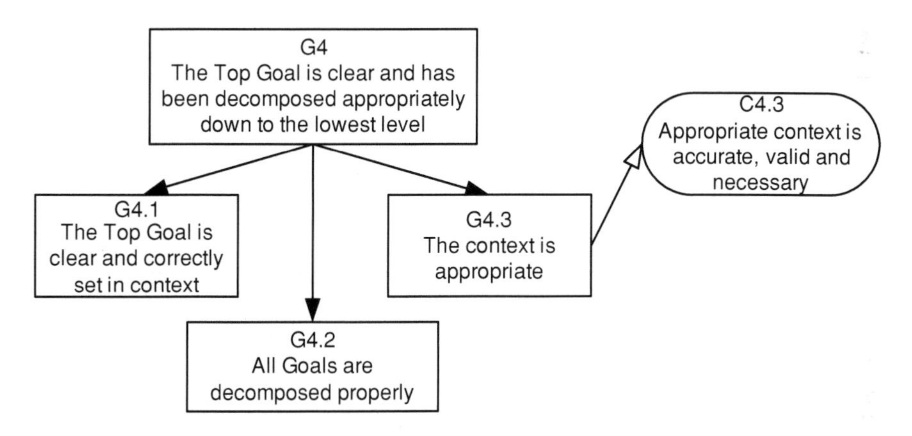

Fig. 9.14 Claim, decomposition and Context are correct

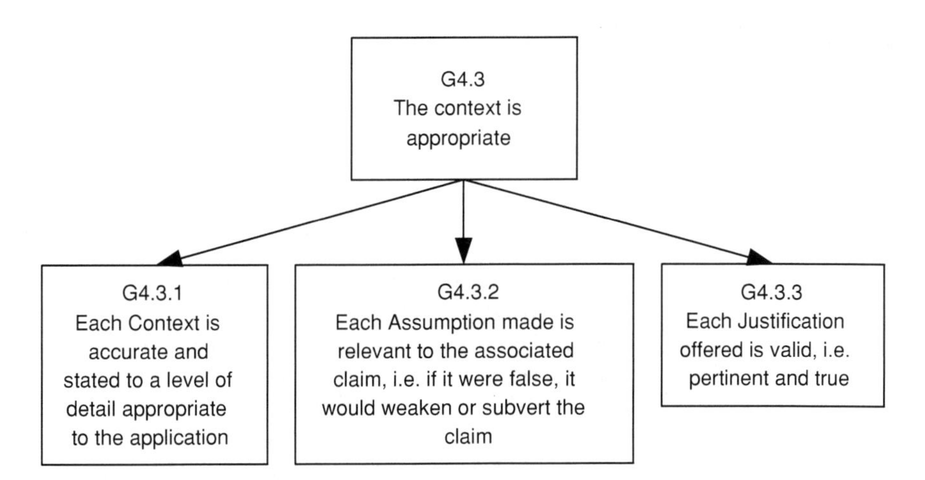

Fig. 9.15 Context is appropriate

In Goal G4.3.1 (Fig. 9.15), the ‘level of detail appropriate to the application’ means check whether there is enough, and check that there is not too much.

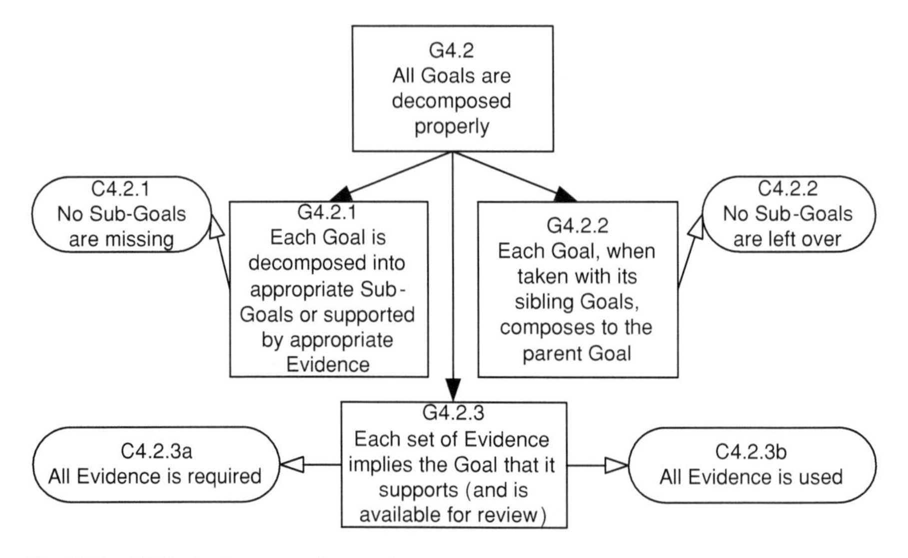

Fig. 9.16 All Goals decomposed properly

9.10 Authors’ Checklist: General

As already noted, authors should ensure that Goals G1, G2 & G3 are fulfilled before sending their arguments out for review.

To generate the checklist, I will re-express a Goal as a requirement to be met by your document and then list the associated questions. To save space, I will set out my questions as bullet points from which you can develop your own checklist(s). When you make your tailored checklist, it may be best to put it into a tabular format, with columns for questions, ticks (i.e. checks) and observations, or just questions and answers if you prefer. It is unnecessary to record the bulleted requirements on the final checklist unless you are going to make it available to others, who may come up with alternative ways of fulfilling some requirements. This is more likely to be the case with the reviewers’ checklist.

Most of the questions require you to check lots of similar things, e.g. “Are all circular elements intended to be Evidence?” If you are easily distracted, or likely to be disturbed, I recommend that you get a highlighting pen and dab each symbol as you check it. As you go along, keep a note of which diagrams need fixing. Only update them when you have finished your checks, as there may well be more than one thing wrong with a diagram; it may transpire that you have to delete a diagram or two; it would have been a waste of time to have updated them already.

9.11 Authors' Checklist: Quality

If you have them, I assume that you will apply your existing checks for correct document template, header & footers, documentary data and configuration management aspects. If you do not have any existing quality checks defined, consider whether you need them and add them to your instantiation of this checklist.

Note that this checklist looks at more than just the presentation of the Goal Structure; the content of the introductory text and that linking the diagrams is also important. If you are presenting an assurance argument, do not put the reader off before you start. I have seen examples that raise expectations and dash them within the first few sentences. "We developed a comprehensive assurance plan early on in the project—but we did not necessarily implement all of it." You need to start with a selling message, not a disclaimer.

- The argument shall be properly presented as legible diagrams.
  - Are the widths and colours of all lines appropriate for ease of viewing, both on screen and in print?
  - Is all text on the diagrams in a clear font and of a size and colour appropriate for ease of viewing, both on screen and in print?
  - Are all words on the diagrams spelled correctly?
  - Is all text on the diagrams grammatical and concise, but without unexplained abbreviations, acronyms or initialisms?
  - If labels are used for the symbols on the diagram, e.g. G3.1, are they clear and unique?
- The text introducing the argument shall be clear; if more than one diagram is presented, the relationships between them shall be clear.
  - Is all text in the document in a clear font and of a size and colour appropriate for ease of viewing, both on screen and in print?
  - Are all words in the document spelled correctly?
  - Is all text linking and/or introducing the diagrams grammatical and concise, but without unexplained abbreviations, acronyms or initialisms?
  - If labels on the diagrams are used for navigation, e.g. "See Section 4.1", are they clear and correct?
- The argument shall be presented as such; not as "The GSN"
  - Is the main heading "The Argument" or the text of the Top Goal (preferred)?
  - Are all document section headings that relate to the argument expressed in meaningful terms, e.g. propositions or parts of the strategy?

9.12 Authors' Checklist: Correct Symbols

This section checks that all symbols used are what they seem and that there are no undefined symbol types.

- All argument elements shall be correctly presented. Note: In the question set below, "enclose" means that all text is inside the symbol boundary.
  - Are all elements intended to be Goals horizontal rectangles?
  - Are all elements that are horizontal rectangles intended to be Goals?
  - Do all Goal symbols enclose a succinct unambiguous claim (which may be true of false) in subject-verb-object form?
  - Are all elements intended to be Strategies horizontal parallelograms?
  - Are all elements that are horizontal parallelograms intended to be Strategies?
  - Do all Strategy symbols enclose a clear statement of rationale, e.g. "Argue that the criteria are necessary and sufficient", or the name of an argument type, e.g. "Independence Argument"?
  - Are all elements intended to be Evidence circular?
  - Are all circular elements intended to be Evidence?
  - Do all Evidence symbols enclose a clear reference to the actual item of evidence (and is that item to hand)?
  - If the actual evidence is only part of an external document, is that part clearly referenced?
- All contextual elements shall be correctly presented
  - Are all elements intended to be Contexts horizontal caplets or rectangles with rounded corners?
  - Are all elements that are horizontal caplets intended to be Contexts?
  - Are all elements that are horizontal rectangles with rounded corners intended to be Contexts?
  - Do all Context symbols enclose either succinct explanatory text, e.g. a definition, or an external reference to supporting material, e.g. an International Standard (and is that material to hand)? Note: Context symbols shall not include the text "In the context of..."
  - Are all elements intended to be Justifications horizontal ellipses with a J?
  - Do all Justification symbols enclose either an axiom or factual statement, a brief argument, or a reference to an external argument (and is that external argument to hand)?
  - Are all elements intended to be Assumptions horizontal ellipses with an A?
  - Do all Assumption symbols enclose a succinct proposition that is to be taken as true? Note: Assumption symbols shall not include the text "It is assumed that..."
  - Are all elliptical symbols labelled with either an A or a J; are all those with a J intended to be Justifications and all those with an A, Assumptions?

- All remaining symbols shall be relationships; lines with a single arrowhead, hollow or solid as appropriate.
  - Having checked all argument elements and contextual elements, are there any symbols left over that are not representing relationships between elements? If so, correct them or delete them.
  - Do all lines at the top of argument elements have solid-headed arrows?
  - Do all lines emerging from the bottom of argument elements have solid-headed arrows at the other end?
  - Do all lines emerging from the sides of argument elements have open-headed arrows at the other end?
  - Are all lines attaching to contextual elements open-headed arrows?

9.13 Authors' Checklist: Correct Relationships

This section checks that all relationships are correctly portrayed.

- All threads of argument shall contain valid argument elements in correct relationships.
  - Does each Goal (other than the Top Goal) have at least one incoming Thread of Argument arrow from a Goal or Strategy?
  - Does each thread of the argument terminate on either a Goal to be Developed or a Goal supported by one or more Evidence symbols?
  - Does each Goal not supported by evidence have either more than one outgoing Thread of Argument arrow (all to Goals, or all to Strategies) or a single such arrow to a Strategy?
  - Does each Strategy have more than one outgoing Thread of Argument arrow; all to Goals?
  - Does each Evidence have at least one incoming Thread of Argument arrow; all from Goals?
- All contextual information shall be presented in correct relationships with argument elements.
  - Does each contextual element have at least one incoming open-headed arrow from a Goal or a Strategy?
  - Does each contextual element only have incoming (open-headed) arrows?
  - If any contextual element has more than one incoming arrow from related Sub-Goals or multiple Strategies, would it be possible to move it to a common point higher in the Goal Structure?
  - If any contextual element has more than one incoming arrow; would it be clearer to repeat the symbol adjacent to each of the associated Goals or Strategies?

9.14 Authors' and Reviewers' Checklist: Correct Argument

Because the author uses the previous parts of the checklist and make appropriate changes before sending the argument out for review, it should be unnecessary for the reviewers to have to use it. In contrast, both reviewers and authors should use the following part of the checklist. The viewpoint of the author should be, "Have I put my argument across clearly and correctly?" The reviewer should be asking, "Do I understand this argument; is it what is needed?"

- The Top Goal shall be clear and correctly set in Context.
  - Is the meaning of the Top Goal clear, and is it what you intended to argue?
  - Is the context in which you argue for the Top Goal clear, correct and not overly constraining?
- All Goals shall be decomposed properly.
  - Is each Goal decomposed into pertinent Sub-Goals (possibly via appropriate Strategy), or directly supported by pertinent (i.e. both appropriate and of correct issue state) Evidence; are any Sub-Goals or Evidence missing?
  - Are any additional Strategies required?
  - Does each set of Sub-Goals readily compose to their parent Goal, such that no Sub-Goals are left over?
  - Does each set of Evidence clearly imply the Goal that it supports (possibly explained by appropriate Justification), such that all Evidence is used and there is none missing, or left over?
  - Is the set of Evidence actually available for review? Is it complete and does it support the argument to the degree required?
  - We should have checked this before, under "Correct Symbols", but this is a good opportunity to check whether the Evidence references are specific enough; when reviewing the Evidence items, did you find the part of the document you needed easily (because it was referenced in the Evidence symbol, or in its Justification) or did you have to search?
- All contextual information shall be appropriate, i.e. accurate, valid and necessary.
  - Is the content of each Context accurate and stated to an appropriate level of detail, i.e. is there sufficient detail, but not too much?
  - Are any additional Contexts required; are any superfluous, or too constraining?
  - Is each Assumption valid and is it relevant to the associated claim, i.e. if that which is assumed were to be false, would it undermine the claim?
  - Are any additional Assumptions required?
  - Is each Justification offered pertinent and true?
  - Are any additional Justifications required?

9.15 Consolidation

Before deriving my checklist questions I said that, when completed, I would compare them with those in a published paper on argument review (Kelly 2007). I will concentrate on the Reviewers' checklist as that is the scope of the paper.

This is a good example of an Assumption that proves to be invalid; from context I had assumed that the paper would be addressing the review of assurance arguments that are presented in GSN. I was wrong, it discusses arguments more generally; for example, it suggests that, if you are given a text-based argument for review, you attempt to draw the structure in GSN or another structured notation. I can, however, look at the bit of the paper that addresses Argument Criticism and Defeat and compare my questions against it.

There are three sub-sections, the first draws out the difference between deductive and inductive arguments (see Chap. 4) and notes that assurance arguments are usually inductive. It asks, in effect, whether the bottom-level Goals and their Evidence are strong enough to support the proposition in the Top Goal; are they sufficient? My review questions are deficient on a number of points when compared with the sufficiency bullet points in the paper; I will address each in turn, paraphrasing the meaning of each:

- Coverage: Check to what extent the argument addresses the Top Goal. This is discharged by my checks of appropriate decomposition into Sub-Goals and the ability to compose each set of Sub-Goals back to its parent Goal.
- Dependency: Check, if you are providing an argument based on two sets of Evidence supporting each other, whether they come from independent sources. I missed this one; however, in mitigation, we have not got to that topic yet in the book (See the Diverse Sensors example in Chap. 10). I suggest in the Questions and Problems sections below that you make a start on your own checklist now, and augment it as you decide which features you need. I have addressed relatively simple arguments using the core GSN here.
- Definition: Check that the context does not over-constrain or under-constrain the argument. This is addressed in my first two questions on contextual information.
- Directness: Check to what extent the Goal decomposition and Evidence directly support the claim being made. For example, if a Goal claims, "It is Safe", does the set of Sub-Goals discuss properties to do with safety, or do they address the problem indirectly through application of "best practice"? I believe that my questions on Sub-Goal composition and Evidence clearly implying the Goal it supports address this topic.
- Relevance: Check how relevant sub-arguments and Evidence are to the Top Goal. This is addressed by the Evidence clearly implying its Goal question and by the check that the Evidence is pertinent, i.e. both appropriate and of correct issue state.
- Robustness: Check how sensitive the argument is to possible changes in the Evidence. For example in Chap. 8, I presented an example where Reliability Block Diagrams had been used to predict a value to compare with the required

target; if it later transpired that the analysis had been overoptimistic, how would it affect the overall argument? I missed this one; I need to augment my review argument with, inter alia, consideration of the pitfalls identified in Chap. 8. I will, as they say, leave that as an exercise to the reader. The intention of this Chapter was to show how you could go about generating checklists for GSN authors and reviewers; the final checklists should be tailored to the type of arguments you need to develop and/or review.

The paper (Kelly 2007) also addresses audit of the Evidence; it reinforces the point made earlier in this chapter that the confidence of the reviewer will be impacted by the number of mistakes they find. Other points made address defeat or undercutting of the argument (see Chap. 13), also the competency of processing and checking personnel and the assurance of tools, which I cover in Chap. 14.

9.16 Questions

Are you going to use my questions as the basis of your own review checklist? If so, you will need to add some questions later if the arguments you review use the generic and/or modular argument symbols that I introduce in Chaps. 11 and 15 respectively.

9.17 Problems

1. I want to break up a large Goal Structure so that I can publish it in a document; can I use the Goal to be Developed symbol to say, "This sub-goal is developed on a later page"?
2. If you are going to develop and present arguments in GSN, make a start on constructing a checklist that you can use when producing a Goal Structure and, in particular, before you release it for review by your "audience".

Reference

Kelly TP (2007) Reviewing assurance arguments—a step-by-step approach. In: Proceedings of a workshop on assurance cases for security—the metrics challenge, Dependable Systems and Networks, July 2007

Chapter 10
A More Interesting Example

Abstract This chapter poses a currently unanswered question; from it, I develop a Top Goal and Context for a Goal Structure. It is a conjecture for which we can argue. By decomposing this Top Goal and its Sub-Goals, we can see where Evidence is available and where it is weak or missing. The Goal Structure enables us to test the argument; we can see where more work needs to be done in support of the conjecture and where we have done sufficient for the current purpose. In this chapter, I use the example to illustrate how an argument may be made more compelling by introducing additional evidence from an independent source.

10.1 Introduction to the Example

When we first looked at contextual information, in Chap. 3, I noted that one would often produce a Goal Structure in response to a question, such as: "Is it safe?" or "Have I done enough?" For this example, I will pose a question that I believe is currently unanswered. I will then state a Goal, assuming the answer is "yes" and develop the upper layers of a Goal Structure. We can then use that to see for what aspects of the topic the argument is weak, i.e. where more work needs to be done in order to answer the original question. In this particular example, "more work" means looking for additional Evidence but, in other examples, it may mean strengthening the argument by modifying it.

Experience of presenting at conferences has shown that whatever example I choose to illustrate a point, there will always be someone in the audience who is familiar with the underlying subject matter and who will say that I have over-simplified, thus missing the point of the example... I originally chose the example of this chapter for an audience with little knowledge of the subject, but who should find it interesting. Now it is in a book, the likelihood of someone in the audience being a specialist in the topic is vastly increased. If you are that person, I apologise

J. Spriggs, *GSN—The Goal Structuring Notation*,
DOI: 10.1007/978-1-4471-2312-5_10,

for my oversimplifications; I just hope that my awe at the scientific and engineering achievements in this topic shows through…

In summary, I'll pose a question, and from it develop the Top Goal of an argument. By decomposing the Top Goal and its Sub-Goals, we will see where Evidence is available and where it is weak. In an engineering context, for example when developing safety or reliability arguments, you would want to identify early on in the project where the Evidence is weak, because there will still be time for extra work to be scheduled in. The converse also holds; you may identify areas where too much is scheduled to be done, you can take the excess out of the plan before the resources are committed.

10.2 The Top Goal

As was shown in Chap. 6, development of the Top Goal is often an iterative process. We go through a number of drafts to zero in on exactly the proposition about which we wish to present an argument and in what context. Often, we will be given a clear question to answer and sometimes (albeit rarely) we get the Top Goal stated as a formal requirement.

In this particular case it took me a couple of drafts to get to the question…

My first draft was: "Is there life on other planets?" This is not really what I wanted to explore, as it suggests Mars or Venus. We can, in principle at least, go to both of them and find out. No, considering our solar system is too parochial.

Second draft: "Is there life on planets that are orbiting stars other than the Sun?" This is much better and it is easy to turn it into a proposition: "There is life on planets that are orbiting stars other than the Sun".

Clearly some contextual information is needed to provide scope by defining "life". My first draft of the meaning of "life" was for it to be an abundant diversity of species, as on Earth. That would be difficult to demonstrate, so I have modified the scope of my argument to keep the quantity of organisms, but not argue for the diversity of species. I have also made my job a bit easier by characterizing the organisms as something that change their environment.

We can also simplify the Goal by using a technical term for "planets that are orbiting stars other than the Sun" and explaining that in another Context. My proposition is thus: "There is life on Exoplanets". It is represented in GSN below, in Fig. 10.1.

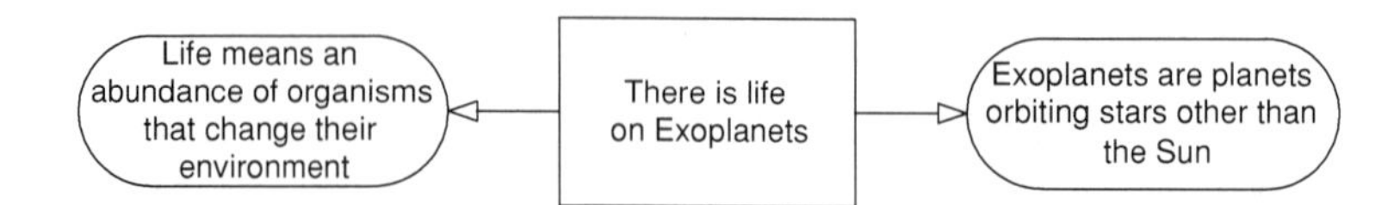

Fig. 10.1 The Top Goal with Context

10.3 A Strategy to Support the Top Goal

Just giving something a fancy name does not make it real, so the first thing I need to do, to persuade you of the truth of my proposition, is to establish the existence of exoplanets. Given that there are such things as exoplanets, I then need to show that the conditions for life can exist on at least a sub-set of them, e.g. there are exoplanets that are neither too hot nor too cold for life.

Having the conditions for life is not sufficient; I also need to persuade you that something is living there in abundance. It therefore seems that I need three sub-goals to support my top goal.

On reflection, there is something else to consider. Because the arguments that you will develop in GSN are meant to persuade, it is sometimes a good idea to include extra material that is not logically necessary, but which responds to a known concern of the audience, or an expected counter-question. Remember the bypass example of Chap. 4? "We need a bypass because A, B, & C … and you tried everything else already". When persuading people, a clever argument is not one of great subtlety and cunning; it is one that addresses the audience's concerns or objections before they have expressed them (of course, if you can do this in a subtle and cunning way, all the better; people do not like to be obviously pre-empted at every step).

In this case I do not know of any such objections for certain, but I can imagine someone saying, "That's all very well, but why has SETI not found anything?" SETI, the Search for Extra Terrestrial Intelligence, is a portmanteau name for various initiatives inspired by the work of, inter alia, Carl Sagan and Frank Drake that are searching for electromagnetic transmissions from distant civilizations (Drake and Sobel 1993). It has yet to produce convincing repeatable results (Fig. 10.2).

About fifty years ago, the number of transmitting civilisations out there was estimated and, basically, if there are that many, we ought to have heard from them by now. The estimate was based on a set of assumptions related by the so-called Drake Equation. I am not requiring intelligence of the life I am arguing for, but it

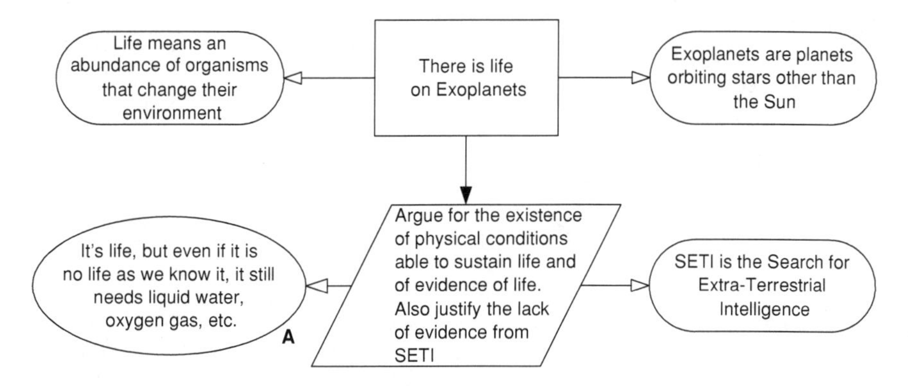

Fig. 10.2 The Top Goal with Strategy and Context

is likely that someone will, in an attempt to undermine my argument, bring up this failure to intercept signals. I anticipate this attack and will refute it in advance.

Note that I have simplified the Strategy by apparently missing out the argument for the existence of exoplanets—I have not missed it out completely; it is implicit in the "physical conditions" clause.

Strategy statements should be succinct; they do not need to explicitly identify each Sub-Goal, but neither should those Sub-Goals be a complete surprise…

10.4 Decomposition of the Top Goal

The Sub-Goals I need to support the Top Goal are thus to:

- Establish the existence of exoplanets;
- Show that the conditions for life can exist on at least a sub-set of exoplanets;
- Show that there is evidence for life out there; and
- Refute an expected attack before it can be mounted

The first level of Goal decomposition is shown in the figure below. Note that I have introduced labels so that I can, for example, just refer in the text to "Goal G3" instead of "the Goal claiming that evidence of life is found in exoplanet atmospheres". This will also make it easier for reviewers to refer to their areas of concern. Note also the flow down of context; in Fig. 10.3, I do not have to say what an exoplanet is at Goal G3, say, because it has already been said at the level above.

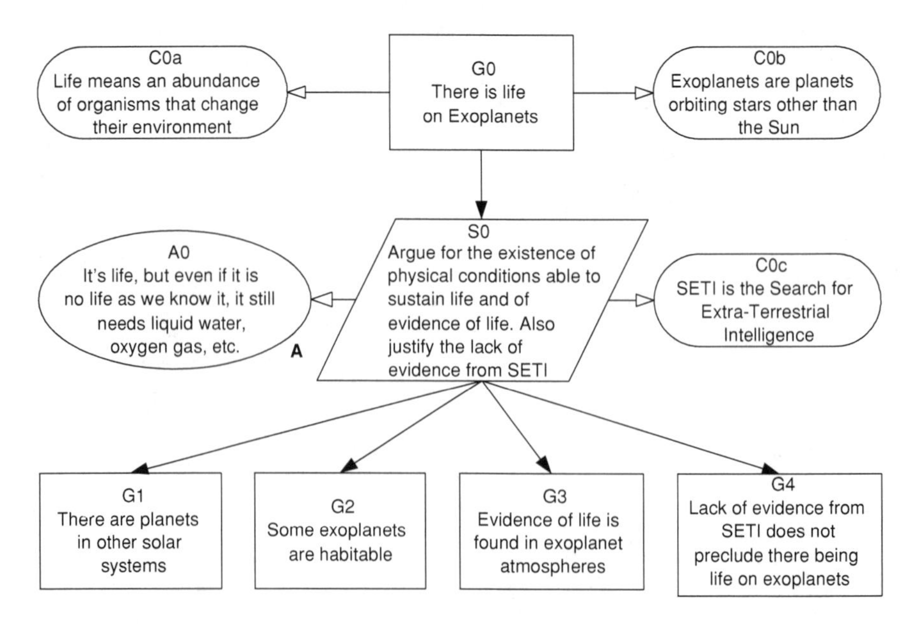

Fig. 10.3 The first level of decomposition

Note: Assumption A0 paraphrases what Spock (played by Leonard Nimoy) said to Captain Kirk (William Shatner) in the *Star Trek* episode called "The Devil in the Dark" (Coon 1967).

10.5 How to Detect Exoplanets?

I assert that there are planets orbiting stars other than our Sun, but how can I convince you of that? You can tell the stars are there because you can see them, when the conditions are right. Surely, planets are too small to see.

To actually see exoplanets probably requires travelling much nearer to them, but we can infer their existence from effects that we can observe from here. I have loosely referred to planets orbiting stars, but in fact the star-planet system rotates about its centre of mass. In general, a star is very much heavier and larger than a planet, so the centre of mass of the system is inside the star and the rotation around it makes it look as if the star is wobbling. We are looking for a wobble of a light-emitting body—this sounds like a job for Doppler detection techniques.

The Doppler shift is the observed frequency of a wave changing as the relative speed of the source and the observer changes. It is usually explained in terms of the changing pitch of the sound of a Police car siren as it approaches, passes and then goes away from the observer, but in fact Professor Doppler's original work was on the coloured light of stars (Doppler 1843). He was interested in binary stars, two stars orbiting each other (actually their common centre of mass), and explained their colour as being due to what has subsequently become known as the Doppler shift. The effect is smaller for a planet causing a star to wobble but, remarkably, it is observable.

Some sub-goals spring to mind, "Doppler sensors are capable of detecting star wobble", "Star wobble suggests the presence of exoplanets" and "Results are available from Doppler sensors"; see Fig. 10.4.

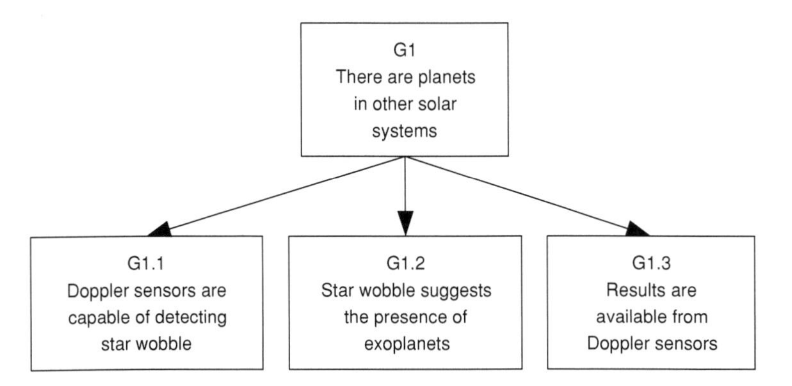

Fig. 10.4 There are planets in other solar systems

Yes, but it is not very convincing, is it? I am claiming that the sensor data show that the star is wobbling, but what if it is just the surface of the star moving in and out? Do stars "breathe" like that? I do not know, but it is not impossible; the 'surface' we are looking at is not solid, it is a light-emitting plasma that behaves as a fluid. There could be resonances in it that, from our perspective at least, produce observable effects similar to those of a planet rotating about it.

I cannot provide extra argument to persuade you that the Doppler measurements come from rotational effects, rather than surface oscillations. So, following the principle of answering counter-questions before they are asked, I need to find an alternative means of detection. It does not matter if there is also some element of uncertainty in this method, as long as it is independent of the Doppler sensor. We can look at the declarations that each sensor makes and only count those that come from both as a genuine exoplanet "sighting".

10.6 An Alternative Means of Detection

The second detection method we shall use is based on eclipses. Technically, it is "transits" we are looking for, that is the passage of a planet in front of its star. It will not give us a total eclipse, but there will be some dimming of the received light. An eclipse of the Sun can be total because of the relative sizes of the Moon and the Sun and the distance from which we observe their conjunction. If we were to see a transit of Venus across the Sun, for example, it would appear as a black dot. Although larger than the Moon, Venus is much further away from us and closer to the Sun. If exoplanets exist, and we measure the brightness of remote stars, we will observe occasional reductions in brightness as they transit. The astonishing thing is that, considering the distances involved, we can actually detect this effect; results are already available. Measurements of brightness are clearly independent of measurements of frequency, so I claim that the two sets of measurements would be independent, see Goal G1.3 in Fig. 10.5.

Changes in brightness can probably also arise from mechanisms other than transits but, when we combine the results with those of the Doppler sensor, there are correlations when both sensors declare a positive sighting. There is also a periodicity to the results, which is expected from an orbiting system.

Note that a simple Justification of Evidence would not be enough here. The currently undeveloped Goals claiming "Results are available…" will have to be supported by rigorous process arguments. I discuss such arguments in Chap. 14.

If we wanted to be really confident, there are other detection techniques that can be used; we could introduce these extra data sources into the mix, but I'll leave it at this for now. The point of this part of the example is that you can increase confidence by using evidence from different sources only if those sources are independent. If they are not independent, you may be basing your argument on data that are misleading due to an effect common to both sources; two sources are not always better than one.

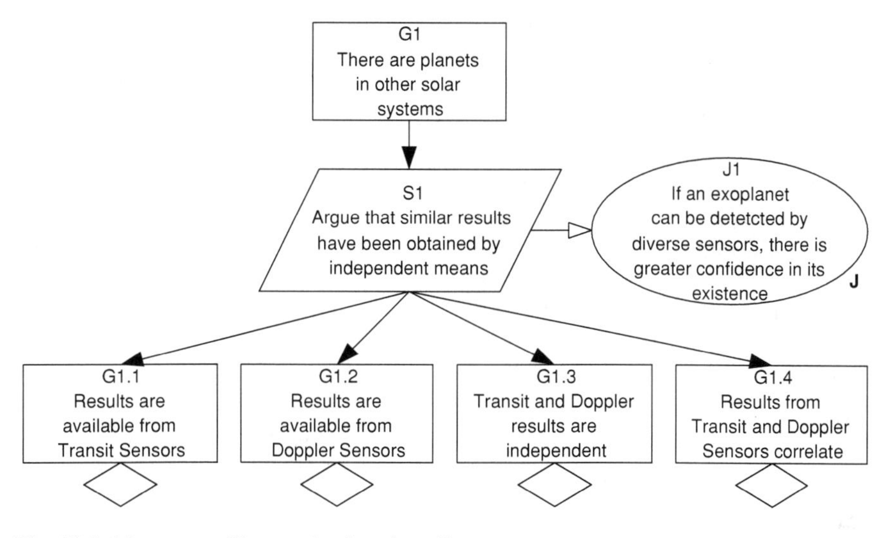

Fig. 10.5 More compelling version based on diverse sensors

I shall return to this part of the example in Chap. 11, "Generic Arguments", to show how you could develop an argument for a general type of sensor, which can then be instantiated for each new type as you deploy it.

Note: Since I first drafted this section, the COROT and Kepler space-based sensors have been deployed to look for exoplanets. With these, and other sensors, now in place, we can expect a lot more detections in the coming years.

10.7 Are Exoplanets Habitable?

Having demonstrated that exoplanets exist, what about the conditions for life? Are the exoplanets that we have detected actually habitable?

This time, I will use what I call an "attrition argument". In the previous 'leg' of the argument, we established, by the correlation of data from multiple independent sources, the existence of a set of exoplanets. Not all of them will be what we are after, so I will now whittle the set down. I claim that, if we discount all those exoplanets that are not habitable in our terms, any that remain will be candidates that we can examine for life signs.

I stated an Assumption with the Strategy for decomposing the Top Goal. It was that the exoplanet-based life forms, whose existence I am arguing for, need the same things in their environment as we do; for example, liquid water and gaseous oxygen. These things only occur under certain conditions.

Take water, for example; on Earth, it typically becomes a solid if the temperature falls below 0°C (zero degrees Celsius), and it typically becomes a gas above 100°C. I say typically, because the precise transition temperatures depend on other environmental effects. Oxygen, on the other hand, has a boiling point of −182.96°C in standard conditions, so it should be fairly easy to argue that it will be a gas in conditions where water is a liquid. Of course, water and oxygen are not the only prerequisites for life; I have just used them as examples of what we are looking for.

The average temperature of a planet's atmosphere (if it has one) is largely dependent upon the distance from its star. If it is close in, a planet will absorb a lot of solar radiation and so will be much hotter than one that is a long way out, where it gets less solar radiation. Close to the star is too hot for life, any surface water would boil away; far out is too cold, any water will freeze.

You may recall the folktale about three bears whose house gets invaded by someone, who, amongst other things, tries the food that has been left out on the table. One bowl is too hot, another too cold, but the third is just right. Similarly, there is a zone around a star that is deemed "just right" for life. It is called the Goldilocks Zone after the intruder in modern versions of The Story Of The Three Bears (Southey 1834). A planet must lie within the Goldilocks Zone in order to sustain liquid water on its surface (and gaseous oxygen in its atmosphere). This is captured in Fig. 10.6.

The first pair of Sub-Goals, G2.1 and G2.2, eliminate from our search the sets of exoplanets that are too hot or too cold. Sub-Goal G2.3 claims that there are places where exoplanets would not be subject to either of these extremes. Sub-Goal G2.4 then claims that we have looked in those places and found planets there. But, what about life signs? That is the focus of Goal G3.

10.8 How to Detect Signs of Life on Exoplanets?

A planet close to its star will be too hot for life to evolve; one too far from the star will be too cold. So we know where we should look for signs of life; in that small proportion of exoplanets that lie within their stars' Goldilocks Zones. But for what signs should we be looking?

Remember, earlier, I made a point of saying that the organisms for which we are looking can change their environment? In particular, their presence should be apparent from an analysis of the exoplanet's atmosphere. Sounds reasonable, and easy to say, but how do we analyse this distant atmosphere?

You know how you can get a spectrum of colours from sunlight by using a prism (or a rainbow by using rain drops)? If we were to use a more precise instrument called a spectrometer, we would see that there are fine black lines in the spectrum. These lines are due to the absorption of the light at frequencies that are characteristic of the material through which the light has passed. Simplistically, absorption lines are related to the permitted energy levels of electrons in the atoms of the material; effects are also observed due to molecular structure. Given precise measurements, we can work out what materials are there, between us and the star.

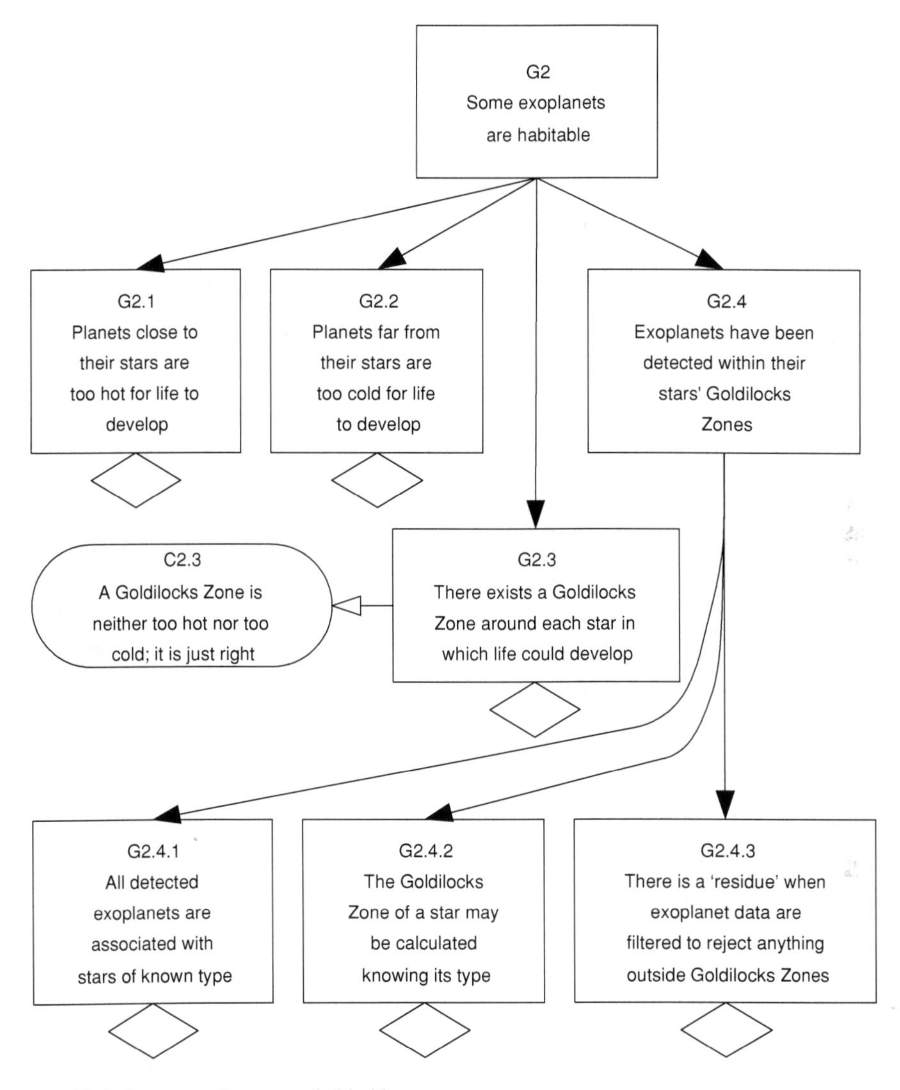

Fig. 10.6 Some exoplanets are habitable

For the Sun, Earth's atmosphere dominates, but with other stars, we can also detect interstellar material.

Pick a star with an exoplanet, examine its light with a high resolution spectrometer, and you get a 'signature' of that star, which is a function of the star and where it is, i.e. what material is in the way. Now integrate the spectrometer with your Transit Sensor (preferably a space-based one so as to get rid of the effects of our atmosphere). You will observe the overall dimming due to a transit plus extra dimming at particular frequencies. Any such additional features identified in the starlight spectrum when the planet passes the star will be due to its atmosphere. This approach does work; for example, methane and water vapour have been detected in exoplanet atmospheres (although not necessarily in Goldilocks Zones).

My decomposition of Goal G3 (Fig. 10.7) is a gross over-simplification; I have (without declaring it) assumed that the exoplanets are like Earth and that we can infer life from the observed proportions of oxygen, water and carbon dioxide in the atmosphere. To convince people of my claims, I will need to produce evidence (supported by a process argument) of atmospheric make-up and justify its importance, i.e. why the observed proportions of gases is deemed significant.

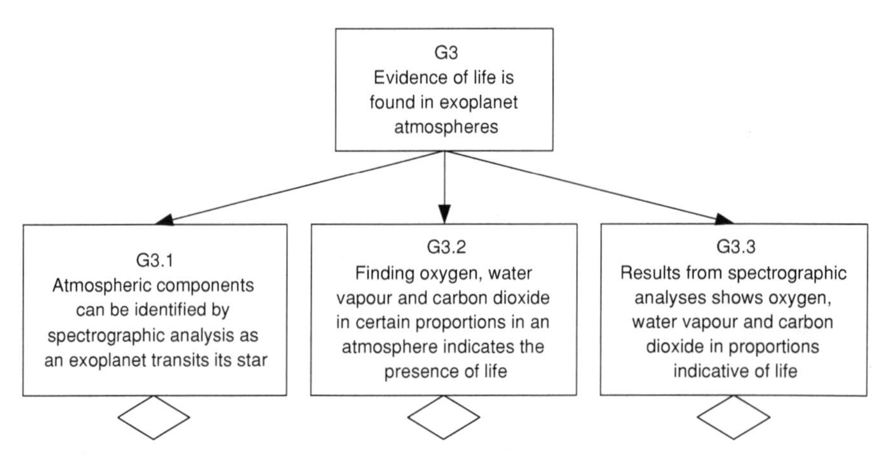

Fig. 10.7 Evidence of life

One approach would be to base the argument for detection of lifesigns on a climate model, which would predict the proportions of gases assuming life were present. I would need to explain and justify my model in the argument but, having done that, I could compare its predictions with what I had measured and make an appropriate claim.

It would be a good idea also to predict the proportions without life, both to show that there is a measurable difference and to give explicit ranges of measurements for Yes, No and Don't Know. If many results for an exoplanet fall into the latter category, it shows that the climate model (or the assumptions that underpin it) needs attention.

In practice, this is where the argument is weak. Few exoplanets have been found in Goldilocks Zones to date; few measurements have been made. I believe that the required results will be obtained once more candidate planets have been discovered. I will address the current lack of evidence in Chap. 12, which looks at what to do if expected evidence is missing.

I also leave decomposition of Goal G4, "Lack of Evidence...", to Chap. 12. It is clearly another missing evidence scenario, but the way we will address it will be different from that of Goal G3. This is because the underlying reason for the evidence not being there is different.

10.9 Questions

- Are you interested in finding out more about exoplanets? If so, a good place to start is The Extrasolar Planets Encyclopaedia, http://exoplanet.eu/. You can find more about the Search for Extra Terrestrial Intelligence from the SETI Institute, http://www.seti.org/.
- In the top-level argument, I anticipated a challenge that may be made, i.e. other experiments have been made (SETI) that should have produced evidence to support my claim, but they have not produced that evidence, suggesting that there is no life out there. As will be shown in Chap. 12, this could be due to an invalid assumption being made. Can you think of other possible situations in which expected evidence may not be forthcoming to support an argument?
- Another potential problem for a practical argument is that, sometimes, you get more evidence than you bargained for. You may get "counter-evidence", i.e. evidence that does not support your claim, rather it is evidence that contradicts it. What, if anything, could be done to save the argument in these circumstances? I shall return to this topic in Chap. 13.
- Have you produced a review checklist, as discussed in Chap. 9? If so, now is a good time to add questions to check that Evidence cited from "independent sources" really was independently generated and does not have some common-cause error.

10.10 Problems

1. Modify the first level decomposition for Goal G3 (see Fig. 10.7) such that, instead of claiming that the proportions of oxygen, water and carbon dioxide are significant, you are going to present predictions from a climate model.
2. Another area in which we measure the relative proportions of things to infer something is in carbon dating. Construct a first level decomposition of the Top Goal claiming, "We can measure the age of an archaeological sample of organic matter with a known accuracy". You need to know that three isotopes of carbon occur naturally; they are known as Carbon 12, 13 and 14 (and, for once, I will allow you to use their conventional symbols, if you know them). When a plant fixes carbon dioxide during photosynthesis it takes up these isotopes in the same proportions as they occur in the atmosphere. Carbon 14 is radioactive, with a half life of 5730 years. After a plant dies, or is eaten, the

amount of Carbon 14 in its remains reduces by beta decay. We can thus assess how long ago the death occurred by comparing the proportion of Carbon 14 in a sample with what we would expect it to be from the atmosphere. We can exploit up other sources of age data to calibrate our measurements to a known accuracy.

References

Coon GL (1967) "The Devil in the Dark" is episode 25 of the original series of Gene Roddenberry's *Star Trek*, first aired in March 1967

Doppler C (1843) Über das farbige Licht der Doppelsterne und einiger anderer Gestirne des Himmels (On the coloured light of the binary stars and some other stars of the heavens). In: Proceedings of the Bohemian Society of Sciences

Drake F, Sobel D (1993) Is anyone out there? Souvenir Press Ltd, London

Southey R (1834) The doctor (Contains the first print version of The Story of the Three Bears)

Chapter 11
Generic Arguments

Abstract The example introduced by the previous chapter contained sub-arguments about different types of sensor device. New sensor types are in development; it would be more efficient if we were to develop a generic argument for any type, and then fill in a copy with the details of new devices as they become available. This chapter uses the sensor example to show how to develop generic goal structures for re-use. It introduces new symbols and text conventions for a Generic Goal (or Goal to be Instantiated), plus some new arrow types representing Multiple Relationship, Optional Relationship and OR Relationship.

11.1 What is a Generic Argument?

In the previous chapter, I started to develop a robust argument for the existence of exoplanets based on measurements made by diverse types of sensor (Fig. 11.1).

If I had based this argument on just one of the data sources, it would have been much less compelling. I noted that, if we want to be even more confident, we could add other types of sensor to the system. It would make it easier for others to 'bring a sensor to the party' if I were to develop an argument for a general type, X, which they can tailor for their particular example. This is a generic argument.

The generic argument is often called a "pattern", presumably after the design patterns used in some engineering disciplines. A pattern in this context is an argument that applies to a class of things, which you can use as the basis of an argument for a specific instance.

Generic arguments can be developed for many reasons. I could, for example, use a generic argument to encourage consistency across different parts of a company; each would use the same basic argument structure and tailor it to their particular needs. Note that provision of a pattern does not remove the responsibility to think about the argument that you are making. The generic argument

J. Spriggs, *GSN—The Goal Structuring Notation*,
DOI: 10.1007/978-1-4471-2312-5_11, © Springer-Verlag London Limited 2012

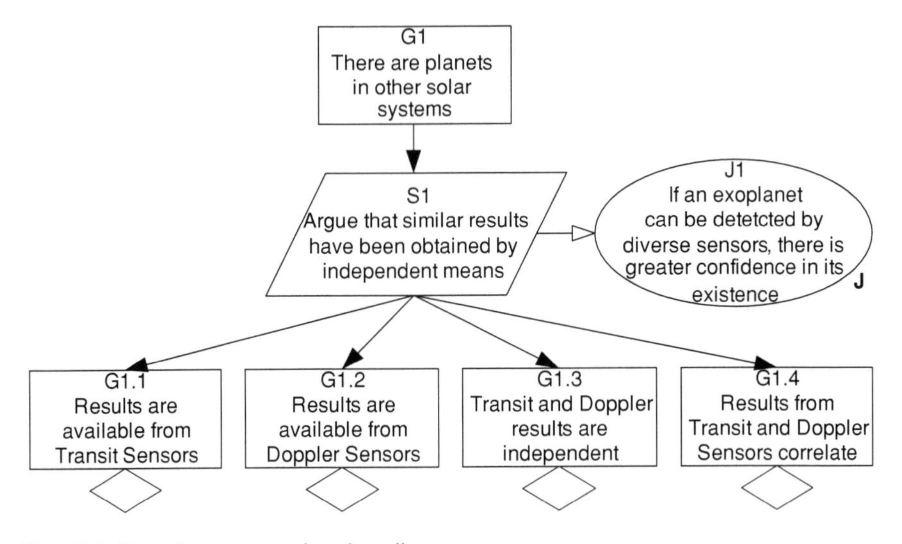

Fig. 11.1 Detection argument based on diverse sensors

standardizes presentation and facilitates review; it does not reduce the task to a filling-in-the-boxes exercise.

11.2 Results are Available from {X} Sensors

The first new symbol we need is in the sub-section heading above. {X} is that part of the subject of the argument which will be specified by the user of the pattern. Of course, it does not have to be an X; I could have claimed that, "Results are Available from {Type Descriptor} Sensors". This latter form is preferred, as it makes it clearer what you are arguing about; in this case it is types of sensor. If you do use an 'unknown' like {X}, add a Context to say what {X} is.

It would be unfair on the reader to expect them to scan your argument for expressions in curly brackets, so there is also an extension to the Goal symbol to highlight a Generic Goal, see Fig. 11.2.

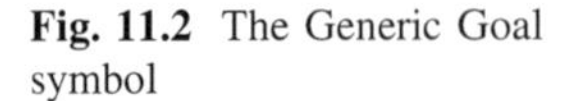

Fig. 11.2 The Generic Goal symbol

In Fig. 11.1 a Sub-Goal similar to this one was shown as undeveloped. The symbol for an undeveloped Generic Goal (Fig. 11.3) is an obvious extension.

Fig. 11.3 The Generic Goal to be Developed symbol

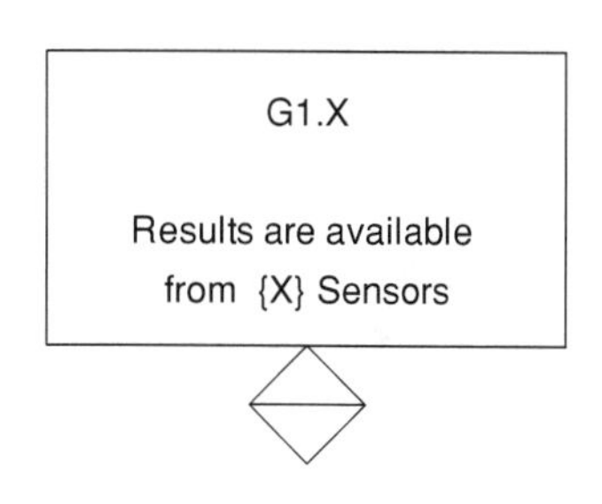

You can interpret the Generic Goal as being a claim that I assert to be true for any member of the set of {X}. When I want to deploy it in an argument for a particular instance, I will replace {X} with the appropriate name and remove the little triangle from the symbol. Producing a particular instance from a Generic Goal is called instantiation, and so you may sometimes see a Generic Goal referred to as a Goal To Be Instantiated or an Uninstantiated Goal.

I will come back to decomposing "Results are Available from {X} Sensors" after introducing some more notation.

11.3 Multiple and Optional Relationships Between Goals

In the sensor example, we need to have more than one instantiation, both a Doppler and a Transit sensor, for example, or we would be back to the earlier and uncompelling single sensor argument of Fig. 10.4. A modified Thread of Argument arrow is used to show if a number of instances is required. Place a blob near the head end of the arrow, as shown at the left of Fig. 11.4 below. This is read as "Supported by zero or more instances of".

Fig. 11.4 Thread of argument modified for multiple relationships

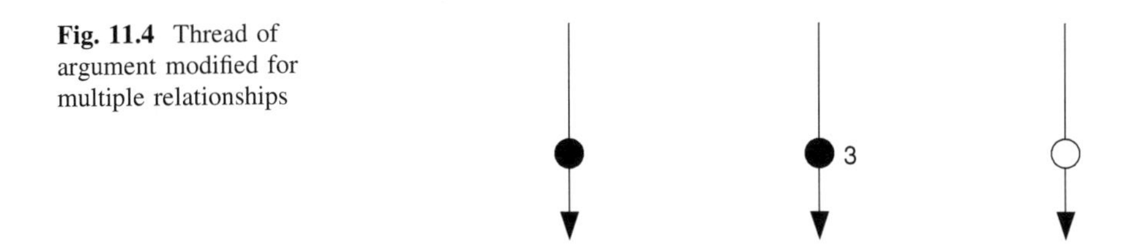

If we want to specify the number of instances, we place a number next to the blob, as in the middle of the figure above; this example requires three instances. This quantity is called the cardinality of the relationship.

Instead of an explicit number, you could put an **n**, say, to indicate that you do not yet know the cardinality; alternatively a simple expression could be used, such as >3, meaning more than three. Sometimes we need to specify either zero or one instance; this is the optional relationship, for which GSN has a special symbol. It is the right hand arrow in Fig. 11.4. To remember this symbol, do not think of it as a white blob on a black one; think O for Option.

An example of an optional relationship is shown in Fig. 11.5; the argument is that entry criteria have been met by the guests. Some need to take prescription medication, so they have an additional criterion that sufficient medication, and supporting documentation, is available.

Note also, in Fig. 11.5, the 'Context to be Instantiated' symbol; the little triangle can be applied to any symbol containing an {X} element.

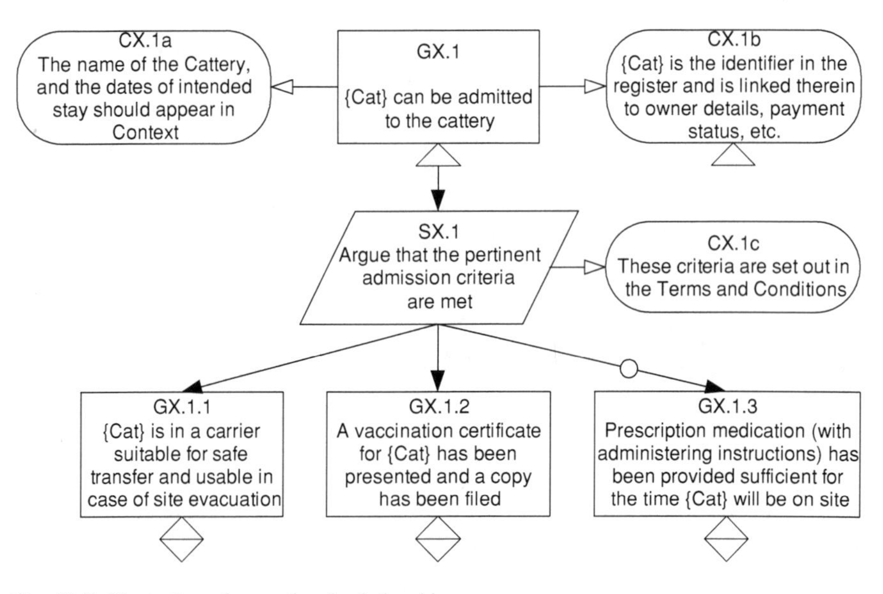

Fig. 11.5 Illustration of an optional relationships

The optional relationship is an 'optional extra' thread of argument. Sometimes we will need to offer a choice. If, rather than assessing cats for admission to a cattery, I were checking candidates for admission to a British university, I would be looking to see which of the many qualification options the applicant holds. Lots of different qualifications are acceptable but, if you were to come along with, say, the right A Level passes from an English examination board, or sufficient Scottish 'Highers', you would not need to offer an appropriate leaving certificate from an Irish educational establishment, or vice versa. Each optional relationship has cardinality zero or one, but I have the additional criterion that they are not all zero. Logically, it is an OR relationship; there is a GSN symbol for this (Fig. 11.6).

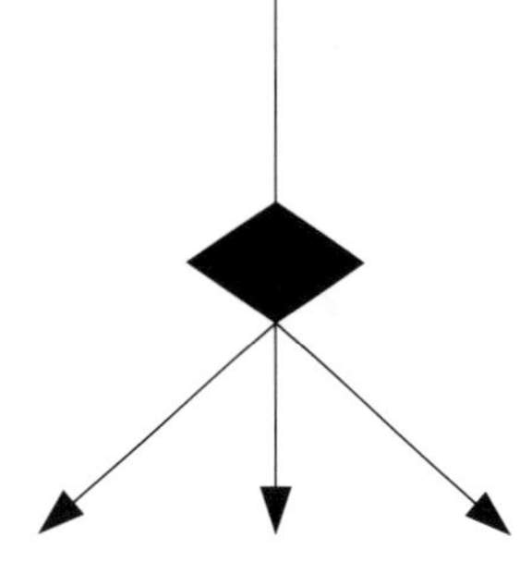

Fig. 11.6 The OR relationship in GSN

Note that you may see an alternative, but similar, symbol used in some published patterns. The original GSN patterns (Kelly1998) used an open diamond, which appeared to be two broken Thread of Argument arrows; see Fig. 11.7, which has one line slightly emphasised to show what I mean about broken arrows.

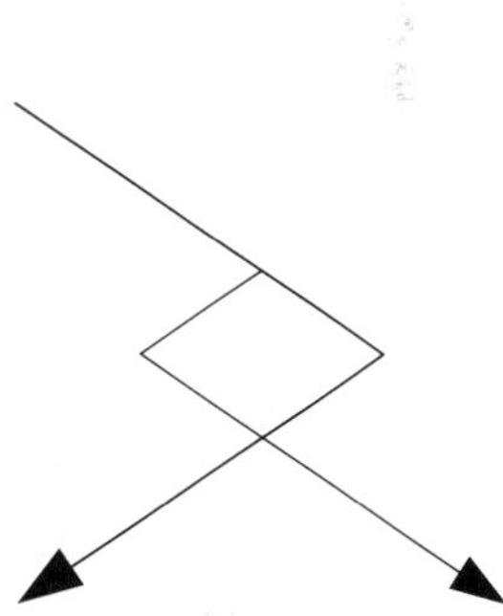

Fig. 11.7 An alternative representation that you may encounter

I have seen potentially confusing examples where the outgoing arrow splits to a number of Sub-Goals giving, for example, an A OR B OR (C AND D) relationship (Kelly 1998). If you need to do this, I recommend that you put a Strategy (Argue C AND D) or an intermediate Goal (C AND D is True) in place such that there is a single symbol at the end of each of the OR arrows.

Figure 11.8 shows the University applicant qualifications example; one out of the options is required.

It does not matter in this example if you have more than one of the options, but sometimes it will. Other times the relationship may be a bit more complex; it many be necessary to choose three of the options, say. This situation is indicated with a number or simple expression in a similar manner to the Multiple Relationship. This is often called an M-out-of-N, or MooN, Relationship (Fig. 11.9).

It is convention to state both M and N even though, usually, N is obvious from the number of outgoing arrows. I say 'usually', because it is valid to have a multiplicity blob on one or more of the outgoing arrows. In many cases, the value of N does not actually matter; it is presentation of arguments for the M that has to be achieved.

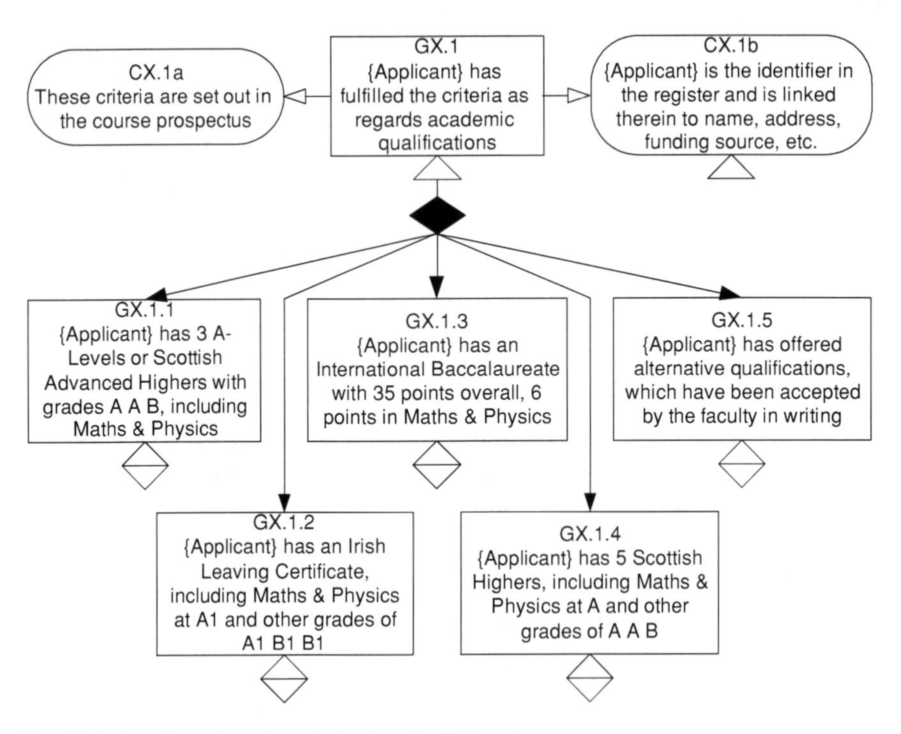
GX.1
{Applicant} has fulfilled the criteria as regards academic qualifications
CX.1a
These criteria are set out in the course prospectus
CX.1b
{Applicant} is the identifier in the register and is linked therein to name, address, funding source, etc.
GX.1.1
{Applicant} has 3 A-Levels or Scottish Advanced Highers with grades A A B, including Maths & Physics
GX.1.3
{Applicant} has an International Baccalaureate with 35 points overall, 6 points in Maths & Physics
GX.1.5
{Applicant} has offered alternative qualifications, which have been accepted by the faculty in writing
GX.1.2
{Applicant} has an Irish Leaving Certificate, including Maths & Physics at A1 and other grades of A1 B1 B1
GX.1.4
{Applicant} has 5 Scottish Highers, including Maths & Physics at A and other grades of A A B

Fig. 11.8 {Applicant} can be admitted to the University

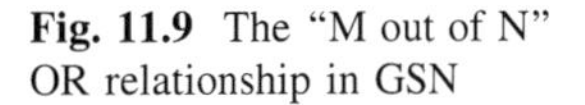

Fig. 11.9 The "M out of N" OR relationship in GSN

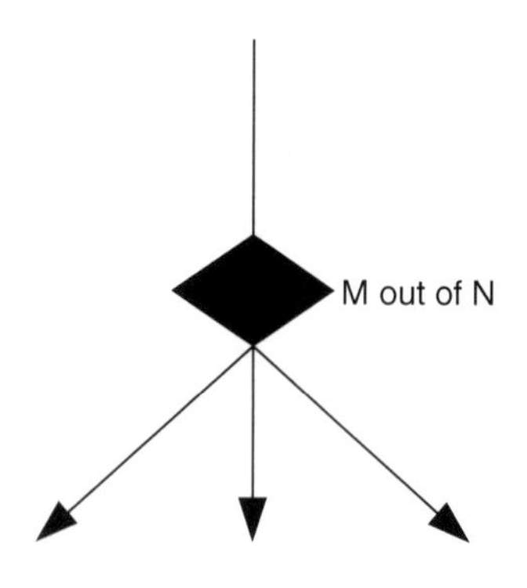
M out of N

11.4 There are Planets in Other Solar Systems

We can take the original decomposition of Goal G1 (Fig. 11.1) and replace the explicit sensor type references with a generic "Results are Available from {X} Sensors" Sub-Goal and a reference to "independent sources", see Fig. 11.10. The generic Sub-Goal can then be decomposed further.

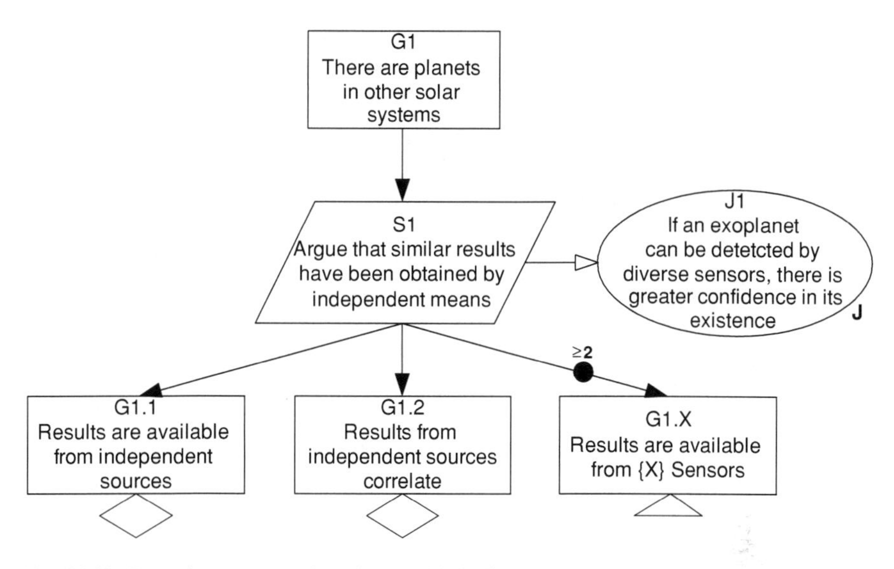

Fig. 11.10 Detection argument based on multiple diverse sensors

My strategy for decomposing the generic Sub-Goal (Fig. 11.11) is to get the provider of a new sensor type to justify that their technology is appropriate, it is sufficiently accurate and that they have got results that can be correlated with those from other types of sensor. They must also justify the independence of their means of detection from those of the other sensor types. In practice the argument below Goal G1.X.4 would get a bit larger each time it is instantiated for a new {X}, as there will be a greater number of types with which to compare.

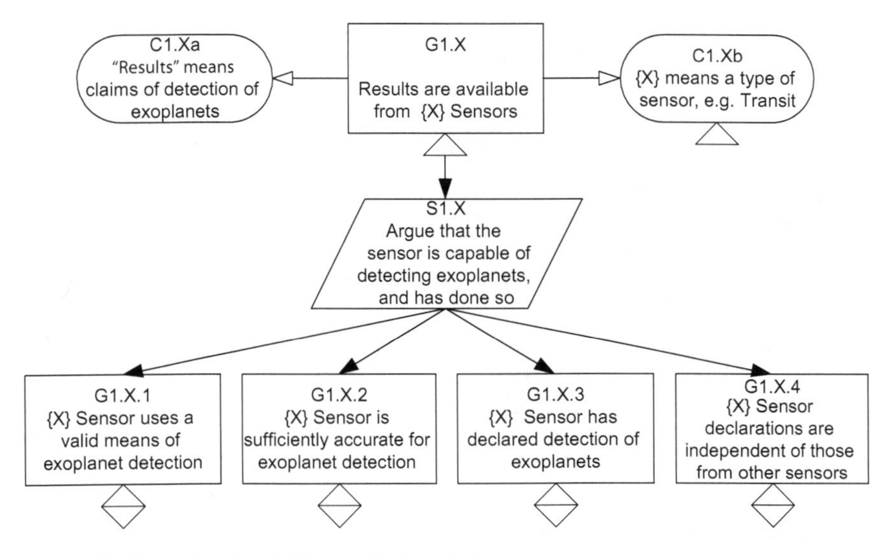

Fig. 11.11 The undeveloped Generic Goal symbol

On examination, the structure is a bit incestuous; one would expect Goal G1.1 to have to appeal to all the instantiations of Goal G1.X.4.

This is an example of where your contractual relationship can affect the argument strategy. If we are all collaborating scientists, maybe operating under a publically-funded research framework programme, we will all swap information about our sensor types and so will be able to construct our version of the decomposition of Goal G1.X.4; Goal G1.1 is unnecessary and can be deleted.

If, on the other hand, we are all commercial organisations working under contract to an Integrating Authority, Goal G1.1 should be retained and the Authority will be responsible for decomposing it. In place of the requirement to decompose Goal G1.X.4, the sensor providers will be required to provide information to the Authority, under a confidentiality agreement, on the workings of their sensor.

11.5 Strategy, Documentary Data and Instantiation Tables

I recommend that you make a point of including explicit Strategy in any pattern you publish for use by others. The purpose of Strategy is to explain the rationale of the decomposition and, combined with Context to explain the scope of the generic argument, should make it a lot easier for the user. This will not be enough in most cases; more information will be required.

You will need to supply some explanatory text with your pattern; I refer to this as documentary data. This should state key information, such as for what you intend the pattern to be used and, in some cases, for what it is not to be used. Any prerequisites or dependencies also need to be stated, for example: "It is assumed that a Hazard Log has been created and properly maintained"; or "This argument depends on an external argument for operator competence".

If you are deploying a pattern for use across an organisation, you should give a point of contact for questions and requests for guidance and also identify an 'owner' to whom formal change requests should be directed.

Perhaps most importantly, you should give instructions and guidelines for use (ideally supported by examples). The clearest way of doing this is to list all the elements that require instantiation, state their purpose and what information is required from the user. There are two approaches that can be used for this, one is to provide annotated lists of all Generic Context, all Generic Goals, etc., the other is to follow the Goal Structure and, where there are generics, discuss each Goal in turn with its contextual information.

In their paper (Kelly 1998), Tim Kelly and John McDermid provide some additional headings that should be considered when preparing the documentary data to go with your pattern.

Once you have done a few of these, you may notice that a lot of the instructions are just for the user to provide data or identify evidence; this looks like a cue for another checklist. I call it an instantiation table; it is another way to present generic arguments. The advantage of this approach is that the user need not worry about the notation, just answer the questions. The argument document is just a copy of, or external reference to, your generic argument, supported by an instantiation table that provides all the required details. For example, I may have an argument claiming compliance with the legislation restricting hazardous substances in products; it would demonstrate how our processes ensure that the amounts of such substances are maintained below the permitted levels. The argument would go into process controls, supplier audits and test methods, etc., whereas all the design authority has to do is fill in the table identifying his or her product and referencing the pertinent Evidence, such as reports of audits against agreed processes.

Such mini-arguments could be referenced from larger assurance arguments via a Justification or as an Evidence.

11.6 Questions

- If you were to develop an argument for reuse in different contexts by the users filling in instantiation tables, do you think it would be clearer just to present it in 'regular' GSN, rather than using Generic Goals, Generic Context, etc.?
- If you work in a regulated industry, do you think that your Regulator would welcome argument patterns for particular parts of the assurance argument? Alternatively, do you think your Regulator may regard agreeing patterns as extra work, although they would make the overall review task easier in the long run?
- Have you produced an author's or reviewer's checklist, as per Chap. 9? If so, add questions to check the use of generic symbols in patterns (and the clear provision of appropriate documentary data).

11.7 Problems

1. You are a system integrator and find that all of your systems have some aspects of their assurance arguments in common; you wish to develop generic arguments to address them. One such relates to the tasks to be performed by the system users. You have a Goal G5 claiming, "User tasks will be performed properly", where proper performance of tasks is defined in the Human Factors Handbook. This is an argument to justify that the system may go into service, so it will ultimately depend (at the lowest level of Sub-Goals) on evidence of specifications, evaluation and training; no evidence from operation, e.g. user performance measures, will be available yet. Suggest a first level of decomposition using {Task} and {User} as appropriate in your Sub-Goal statements; there must be at least one of each for the argument to be instantiated. Note that "first level of decomposition" means to go down to the first line of Sub-Goals

(or two lines, if you find that there are rather a lot and you wish to group them using Goals rather than Strategies).

2. In my discussion of Fig. 11.11, in the context of Fig. 11.10, I noted that the argument should be restructured, but how one would do it depends on your contractual relationship. Present a version of the argument assuming that all the sensor suppliers are commercial organisations working under contract to an Integrating Authority.
3. In my discussion of Fig. 11.11, in the context of Fig. 11.10, I noted that the argument should be restructured, but how one would do it depends on your contractual relationship. Present a version of the argument assuming that the sensor suppliers and the system integrators all collaborating scientists able to exchange information freely without having to worry about proprietary data or Intellectual Priority Rights.

Reference

Kelly T, McDermid J (1998) Safety case patterns—reusing successful arguments. In: Proceedings of IEE colloquium on understanding patterns and their application to system engineering (http://www-users.cs.york.ac.uk/ ~ tpk/)

Chapter 12
Missing Evidence

Abstract In Chap. 10, I noted that there was insufficient evidence to support one of my claims. I also noted that evidence from a famous scientific experiment should be available to support the argument, but it is not; the evidence is missing. This chapter examines what you can do in general if the expected evidence does not materialise, and recovers the situation for the example.

12.1 Why is the Evidence Missing?

Often we construct arguments and then go out and seek evidence that we expect to be there to support the lowest level Sub-Goals. Sometimes we are surprised; the evidence is not there! There are a number of reasons for this to occur. The best strategy for dealing with any particular case depends on what the reason is. So, as already asked in the Questions to Chap. 10, why should there be no evidence?

The obvious answer is that there is no evidence to support your claim because your claim is in fact false. If you suspect that this may be the case, you can confirm it by going out to look for evidence supporting the converse of your claim. Such evidence is known as counter-evidence to your claim and is addressed in Chap. 13. Note that the absence of counter-evidence can be used to support your claim, but it must be backing up some other evidence. It is not compelling to claim the truth of something only on the basis of not having anything that contradicts it. However, it can be compelling to present evidence in support of a claim plus evidence of having actively, but unsuccessfully, sought counter-evidence.

Lack of evidence in support of a claim does not necessarily mean that the claim is false; you may just be looking in the wrong place, or at the wrong time. It is quite common to argue for something based on, for example, a draft report from an accredited external auditor, and finish the argument before having received the formally issued version of that report. The best strategy to adopt here is to get your

J. Spriggs, *GSN—The Goal Structuring Notation*,
DOI: 10.1007/978-1-4471-2312-5_12, © Springer-Verlag London Limited 2012

argument reviewed and conditionally accepted, i.e. acceptance is conditional on receipt of a favourable report. If you follow this approach, I recommend that you change your Sub-Goals to depend upon the audit report and upon completion of any actions arising therefrom. This is just in case they add something (or you have yet to complete what was suggested in the draft).

The dearth of evidence for life obtained from spectrographic analysis of exoplanet atmospheres that we encountered in Chap. 10 is of this type. I still believe the claim is true, it is just that we have not yet found many exoplanets in Goldilocks Zones from which we can collect the data. We need to carry on the search to get more data. In Fig. 12.1, this is made explicit with an Assumption.

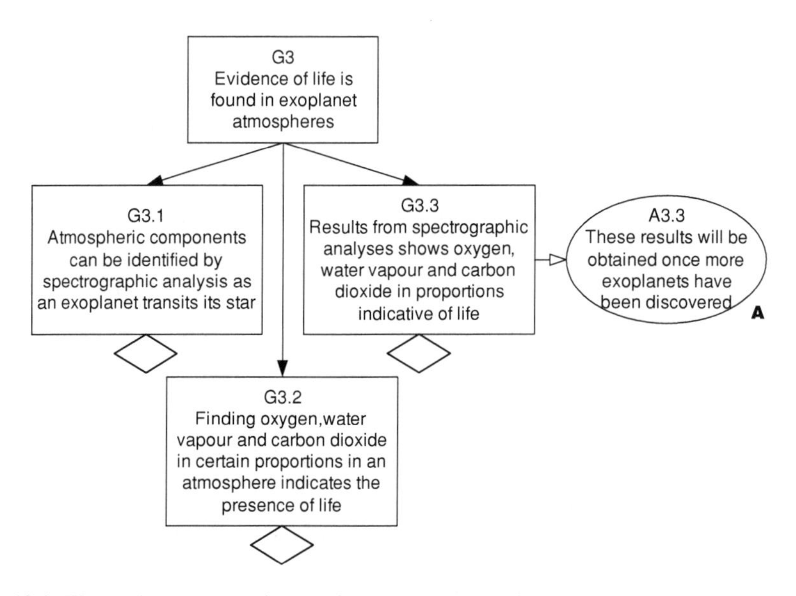

Fig. 12.1 We need to carry on the search

Do not hide the fact that you cannot find sufficient evidence to support a claim, however tempting that may be. Instead, you should use the lack of evidence to direct efforts to strengthen the overall argument, either by actively going out to get the evidence you believe is out there, or by beefing up other parts of the argument so that the lack of evidence is of less consequence overall.

12.2 Were You Mistaken in Making the Claim?

There is a third alternative; there is no evidence to support your claim because your claim is mistaken. It is a reasonable claim, but is false because it was made in the context of an implicit assumption that turns out to be invalid.

This seems rather a fine distinction, mistaken as opposed to false, but the situation often arises in an engineering context and is usually recoverable. Perhaps the most common example of this is the claim, "The equipment passed all its tests" (with suitable inherited context to define the equipment, and local context to refer out to the set of tests). You go in search of the evidence and find that, due to lack of time, not all tests have been run. Note that this is not the case wherein tests have failed, which is addressed in Chap. 13, but rather where the testers did not bother to run all the tests. The author of the argument (Fig. 12.2) was mistaken in assuming that all tests would be at least attempted.

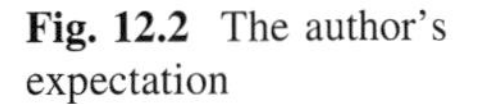

Fig. 12.2 The author's expectation

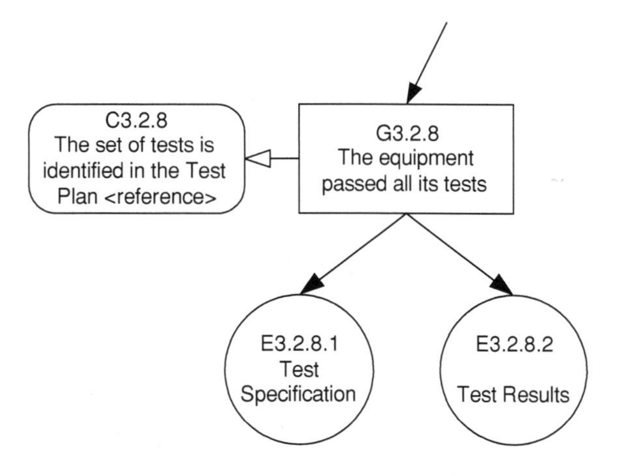

One way out, of course, is to run all the tests and finish late; but professional testers would not have been so cavalier as to just run the first one hundred tests, say; they would have had some rationale for the choice of the tests to run and those to defer. Where there's rationale, there's argument. Can we use that rationale as Justification to save the argument? Often we can; but there is a pitfall to watch out for.

For example, they may argue that, "The first use of this equipment will be in Location A, so the special functions defined for Location B are not yet required; we do not need to test those". Beware, especially if this is in the context of a safety assurance argument; how can you be sure that these extra functions, which are untested, and so potentially faulty, can not in some way adversely impact the operation at Location A? They are not needed at Location A, but how do you know they will not go off unexpectedly, as it were? See Fig. 12.3 for one way out of this.

Note that in many cases it will be easier (quicker and cheaper) to run the whole set of tests than it will be to provide a compelling "non-interference" argument for something that is untested. In Fig. 12.3, I have cheated by removing the untested bits.

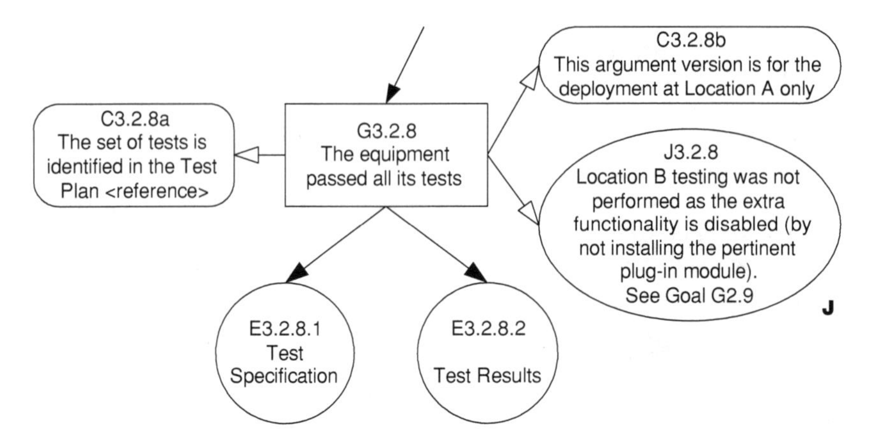

Fig. 12.3 The argument de-scoped for initial deployment

12.3 Summary of Strategies

We have considered three reasons as to why evidence may not be forthcoming:

- The claim may be false: Seek counter-evidence. If you find it, see Chap. 13.
- The evidence exists, but you looked in the wrong places or at the wrong time: Get conditional acceptance for your argument and go out to find the evidence with which to finish the job.
- The claim is true in an assumed context, but the underlying assumption is invalid: Modify, or "descope", the argument to allow for the invalid assumption. You could alternatively do the work necessary to make the assumption valid.

This last alternative, an invalid assumption in the context of the claim, may explain the unsatisfied expectation of receiving signals from extra-terrestrial intelligences. In Chap. 10, I mentioned that evidence for life outside our solar system was expected from a project called SETI, the Search for Extra-Terrestrial Intelligence, but it did not materialise (Drake and Sobel 1993). This expectation was based upon the Drake Equation, also known as the Green Bank Equation, which estimates the number of planets in our galaxy with civilizations that generate radio waves.

12.4 The Drake Equation

The Drake Equation is named after Dr Frank Drake, an American astronomer. He developed it in 1961 when preparing for a conference held at Green Bank, the United States National Radio Astronomy Observatory, in West Virginia. The conference was to discuss the possibility of detecting intelligent life outside of the planet Earth. There is a plaque at Green Bank commemorating the event; it is in the room that once held the blackboard upon which the equation was first publicly written. So as to make it more legible in print, I present the equation on a pretend whiteboard, rather than a blackboard; see Fig. 12.4. The equation has, of course, been developed and expressed differently in the intervening years; my presentation of it here is derived from that on the plaque.

The non-receipt of signals expected from extra-terrestrial intelligences could be due to an invalid assumption underlying the expectation. Can we identify it and recover the argument? Let us consider the Drake Equation. In Fig. 12.4, I have used the same convention as Drake; things written next to each other without an operator are to be multiplied together.

The equation gives us a value for N, the number of civilizations in our galaxy with which we might hope to be able to communicate. There is a very high likelihood of there being inhabited planets outside our galaxy but, if we can find what we are looking for closer to home, our original question, "Is there life on planets that are orbiting stars other than the Sun?" is answered.

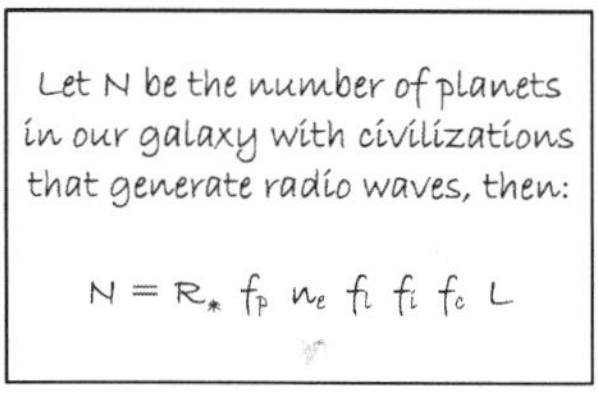

Fig. 12.4 Dr Drake's whiteboard

Let N be the number of planets in our galaxy with civilizations that generate radio waves, then:

$$N = R_* f_p n_e f_l f_i f_c L$$

The other parameters are as follows (with the values estimated by Drake and his colleagues given in parentheses).

- R_* is the average rate of star formation in our galaxy (ten per year)
- f_p is the fraction of those stars that have planets (one half)
- n_e is the average number of planets that can potentially support life per star that has planets (two)
- f_ℓ is the fraction of the above that actually go on to develop life at some point (one, i.e. all of them)
- f_i is the fraction of the above that actually go on to develop intelligent life (one per cent)
- f_c is the fraction of civilizations that develop a technology that releases detectable signs of their existence into space, i.e. radio waves (one per cent)
- L is the length of time such civilizations release detectable signals into space (ten thousand years)

Some of the numbers need to be modified in the light of observations and more-detailed analyses. For example, the average rate of star formation is now estimated at seven per year.

I may have been generous calling the parameter values estimates. The value of L is really a guess; we have no data upon which to base an estimate. Ancient Earth civilizations lasted much less than ten thousand years, would they have survived much longer if they were to have had radio technology? At the start of the nineteen sixties, there had been great strides in technology; we had high power radio and television transmitters in service and no reason to believe that they would be withdrawn. It was expected that they would become "bigger and better" and, looking from outside, the broadcast signals would be observable. Development has not been as expected; we now have more low-powered transmissions, much communication is achieved via cables or point-to-point directional radio links; less will be observable from outside. High power radiated broadcasts will be limited as technology develops; the length of time civilizations release detectable signals into space may well be much less than ten thousand years. Two hundred years may be generous (unless they want to be found, and transmit greetings).

Drake and his colleagues got the answer that N is ten; my guess for L suggests that N may be less than one. Note, however, that in our search for life the parameters f_i and f_c are not relevant, so the number of exoplanets with life in this galaxy could be two thousand. It's life, Jim, but they do not have the means of broadcasting re-runs of *Star Trek*. Figure 12.5 gives a decomposition of Goal G4.

Note that I have imported Contexts from further up the Goal Structure, in case you missed them in Chap. 10. You can tell which they are from the numbering.

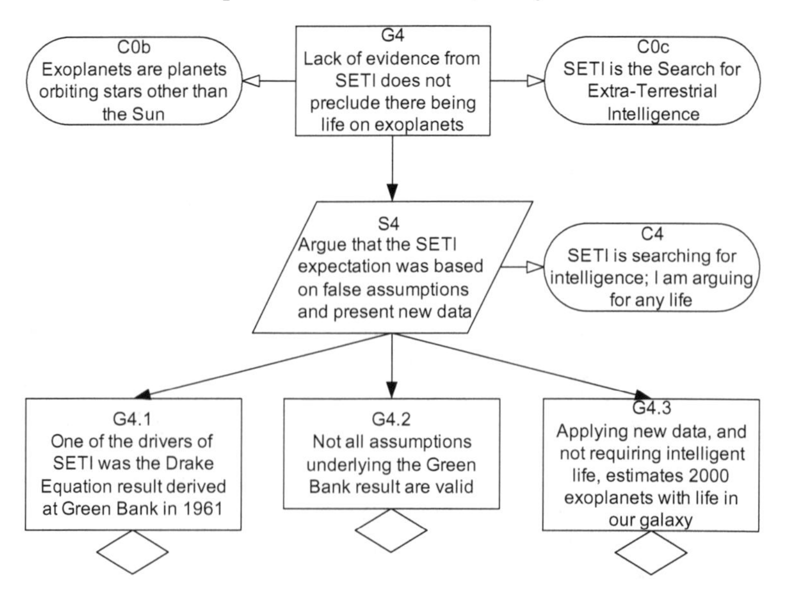

Fig. 12.5 Lack of evidence from SETI is not a problem

12.5 Potential Pitfalls

Be careful, when seeking the evidence that you need, that you are not selecting evidence that only weakly supports your claim whilst ignoring, possibly stronger, evidence that contradicts it. Scott Plous, Professor of Psychology at the Wesleyan University in Middletown, Connecticut, called this "Confirmation Bias". It is defined as a tendency for people to favour information that confirms their preconceptions or hypotheses regardless of whether the information is true (Plous 1993). It is good practice to be aware of this tendency, we all have it, and so we should actively seek out evidence contrary to our claims—a lack of such evidence can strengthen the argument, see Chap. 13.

Another tendency to watch for is the desire to modify the evidence to fit your argument because, sometimes, evidence is not missing, it is just not quite what was expected. For example, a certain UK credit card provider will accept your claim that you are who you say you are if you show them a particular form from the Benefits Agency; a specific document from Her Majesty's Revenue and Customs (but only if dated in a particular period); or a recent statement of account from your bank that shows your surname, initials and title (Mr, Mrs, Dr, etc.) and a payment from your employer. I do not have the Benefits Agency document. I do have that from Her Majesty's Revenue and Customs, but it is earlier than the specified date. So, my Evidence will have to be the bank statement; but my bank does not call me Mister, instead they put Esquire after my name, which the credit card provider does not accept as a title. They suggested that I get my bank to change the form of address they use on my account statements.

Personally, I would not find a piece of evidence that has been tailored specially to fit a claim as compelling as something genuine, from the real world, that does not quite fit. If you do not want to use a Justification to explain the mismatch, change your argument; do not change the evidence.

12.6 Questions

- In this chapter we have looked at possible reasons as to why evidence may not be forthcoming to support a Goal. We observed that, in many cases, the argument can be saved (with some extra work on our part). Chapter 13 considers when evidence is found that refutes the claim of one of our Goals. Can you think how one may save the argument in this case? Can it be saved at all?
- Have you produced a review checklist, as suggested in Chap. 9? If so, add questions about missing Evidence: Has the appropriate strategy been used? Is the argument compelling?

12.7 Problems

Related to the 'no Evidence because of an invalid assumption' case is the invalidation of Evidence that you have already carefully collected. This happens lamentably often in engineering, usually as a misplaced attempt to cut costs. For example, I may justify the use of a computer network component because we successfully used one in an equivalent environment, we have captured all the pertinent network failure logs and analysed them to show that the component worked as required throughout. It transpires that a different component has been procured, deployed in the new system and used throughout testing. The system is ready to go into service as soon as you deliver your updated assurance argument (no pressure). Figure 12.6 shows your original argument for the planned N8102 component; you have the Evidence to hand, but you have just found out that they installed a type N8112 at the sales office. Can you save the day by changing the argument to reflect the deployed N8112 component?

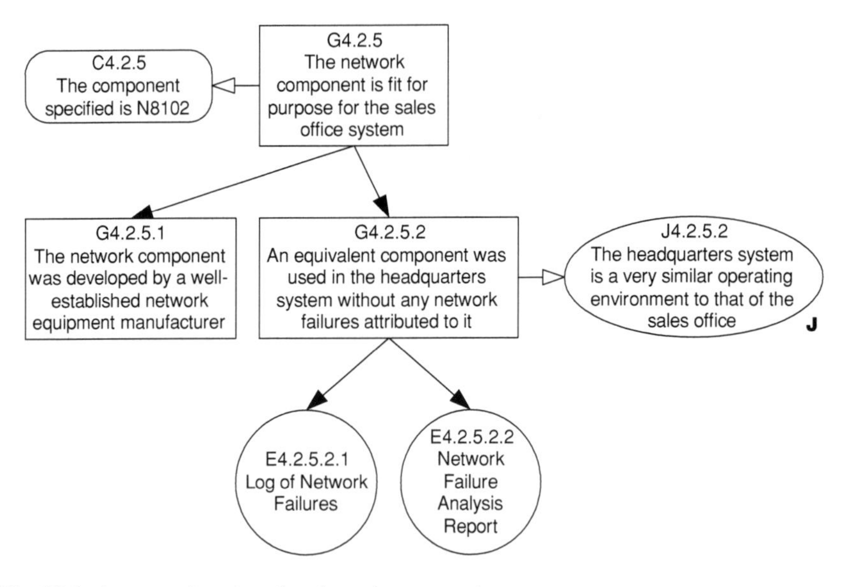

Fig. 12.6 Argument based on the planned component

References

Drake F, Sobel D (1993) Is anyone out there? Souvenir Press Ltd, London

Plous S (1993) The psychology of judgment and decision making. McGraw-Hill, New York

Chapter 13
Counter-Evidence

Abstract Chapter 12 looked at the problem of not finding evidence that you need to support your argument. A potentially worse situation is when evidence is presented that refutes your argument. This chapter suggests strategies for coping with such "counter-evidence".

13.1 What is Counter-Evidence?

In the previous chapter we considered the absence of evidence supporting a claim, and observed that this does not necessarily mean that the claim is false. In this chapter, I will address a potentially more worrying situation; one in which we find actual evidence that indicates our claim is false.

For example, based on the fossil record, nineteenth century palaeontologists claimed that the Coelacanth, a type (more properly, an order) of fish, was extinct and had been for 65 million years, or so. Then, in the 1930s, someone caught one off the coast of South Africa and there was a museum curator on the dock who took charge of it; she showed it to a specialist, who recognised it for what it was... (Smith 1956).

Surely, an argument cannot survive such a blow but, in Chap. 12, we looked at cases where lack of evidence did suggest that a claim may be false, or at least mistaken, and we managed to recover from those, albeit with extra work. Can that be done here?

One of the scenarios considered in Chap. 12 was where the claim is true in an assumed context, but it transpires that the assumption is invalid. We had the case where we claimed that all tests of some piece of equipment had been successful, but found that not all tests had been run. The solution was either to run the tests, or to justify that it was in some sense acceptable to have missed some tests out. In general, it is easier just to run the tests.

J. Spriggs, *GSN—The Goal Structuring Notation*,
DOI: 10.1007/978-1-4471-2312-5_13,

Consider now the related case where all the tests had been run, but some had failed. The invalid assumption being that all tests would be run successfully. Will the same solution as before work? No—run all the tests again and the same ones will fail. If a different set were to fail the second time, you would be in an even worse mess, but that is out of the scope of this book. What about the Justification option; can we ever justify that it is alright to proceed even if some tests have failed? It does depend on what failed; if, for example, running Test 57 caused the equipment to burst into flames, you have a significant problem. Alternatively, if a test resulted in an indicator showing yellow, when it should have been amber, can you live with that for now?

In summary, counter-evidence is not just evidence that does not support your Goal; it is evidence that contradicts it. Do not attempt to suppress counter-evidence; seek to include it, either by modifying your argument to encompass it, or by showing that it does not adversely impact your overall claim, your Top Goal.

13.2 Justification

If a single test has failed and it is a straightforward repeatable failure, like the indicator colour example, the justification that it is acceptable can go in the argument as an additional sub-goal. Figure 13.1 is a repeat of Fig. 12.2, where it introduced the missing evidence example in which not all tests had been run.

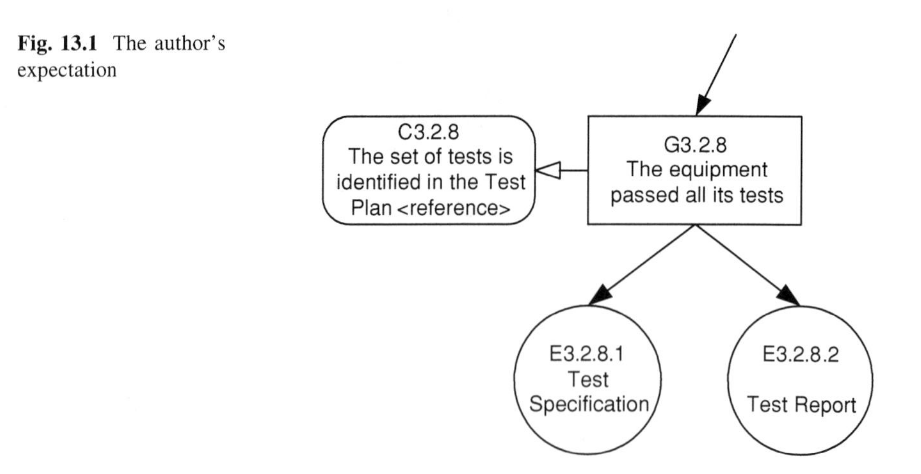

Fig. 13.1 The author's expectation

The scenario here is all tests were run and all passed, except for the indicator colour one, Test 103 say. I have to modify the argument to account for this. I have done this in Fig. 13.2 overleaf; I do not explicitly justify it saying, "it is alright because…", rather I say what I have done about the test failure.

In this case an indicator is not the intended colour, so we have to tell the users, by modifying the User Manual and the material used for their training.

For something like this, it is probably sufficient to supply an erratum slip for the manual noting that, if you have Version 1 of the equipment, the amber indicator is actually yellow. When you have fixed the indicator, just withdraw the erratum (and brief the users, telling them that they now have Version 2, in which the amber indicator lives up to its name).

Usually, an argument like this will be produced for a Customer with whom you have a contractual agreement. In such cases, a procedure will exist for getting Customer agreement for deviations from contracted requirements; I have assumed that the Evidence from this is a formally signed 'Waiver'.

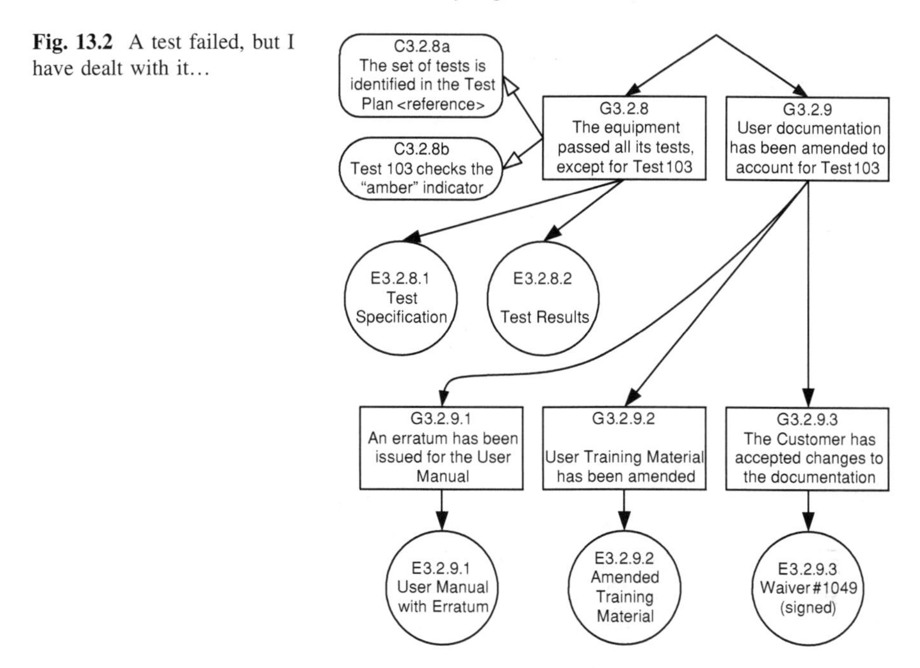

Fig. 13.2 A test failed, but I have dealt with it...

See how the ripples can spread out from what is a simple failure, which was probably caused by fitting the wrong component (or typing the wrong number in a display driver program). Would it not be easier just to fix it?

A more complex failure may mean that some expected function of the equipment is limited in some way, or even missing. You have a shortfall in meeting the functional requirements; a deviation from what was intended. This may still be acceptable, as long as you declare it to the user (who may require you to fix it as soon as is practicable). Rather than provide an erratum for the manual in these circumstances, you will probably need to provide a temporary operating procedure for use until the problem is fixed.

If there are a number of failures, it can get messy to have them all addressed in the argument. Your argument tree structure is marred by a burst of detail, like a mistletoe infestation on a stately Oak tree. You may wish to produce a Test Summary Report that lists, explains and justifies all the anomalies; such a report would appear as Evidence. This is not very 'transparent'; someone reading your argument would not know whether you had two failures to declare, or 200. We need to add some contextual information; we could use a Justification, as in Fig. 13.3.

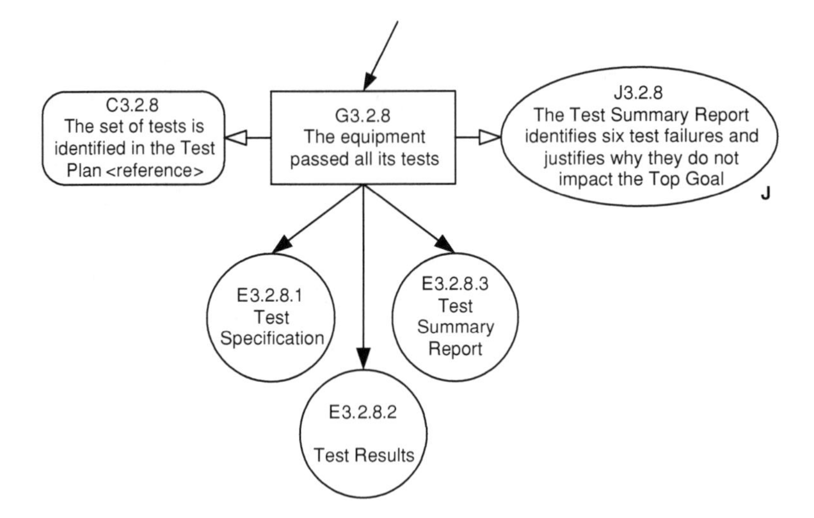

Fig. 13.3 Declaring the size of the shortfall

Perhaps a better option would be to argue each one separately, by slightly abusing the Generic Goal notation as shown in Fig. 13.4. You can then explore each failure in a sub-argument, which will be an instantiation and decomposition of the Generic Goal. I have replaced the Justification with a new Generic Sub-Goal and a Context. The multiple relationship arrow emphasises that there were six failures and that we are addressing all of them individually.

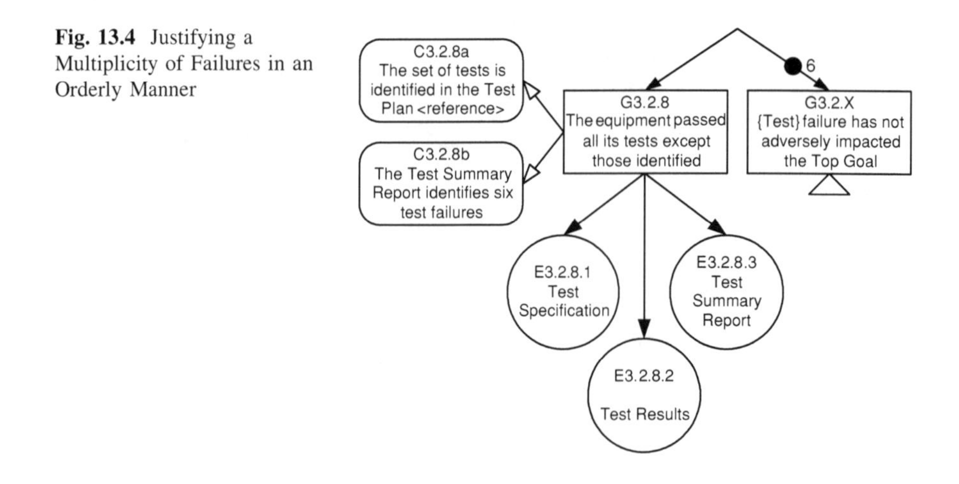

Fig. 13.4 Justifying a Multiplicity of Failures in an Orderly Manner

13.3 Normalisation of Deviance

A pitfall to watch out for here is what Diane Vaughan, Professor of Sociology (also of International and Public Affairs) at Columbia University in New York, called "Normalisation of Deviance" (Vaughan 1996).

The first time you allow a piece equipment into service by adequately justifying that the test failures are acceptable, it will be an uphill struggle getting your six deviations past the Project Manager and all those who have to sign-off the project. Next time around, it may be 66 deviations, but it will be easier to get through. There were no adverse consequences last time, why should there be next? When it gets to 666 deviations, the Project Manager will be trying to convince you that "it's alright". Just say no—that way disaster lies; literally in some cases.

The requirements and quality standards are there for a purpose, why should it be acceptable to erode them over time as your success rate falls? In some organisations this erosion may even be rewarded; a complete normalisation of deviance. The Project Manager who is told, "Your shortcut resulted in the project delivering on time; have a bonus" is not likely to enforce standards next time around either.

13.4 Counter-Evidence is in the Eye of the Beholder

You can use counter-evidence to your advantage; if you are arguing against something, or someone, find counter-evidence to their claims in addition to evidence to support yours, see Fig. 13.5.

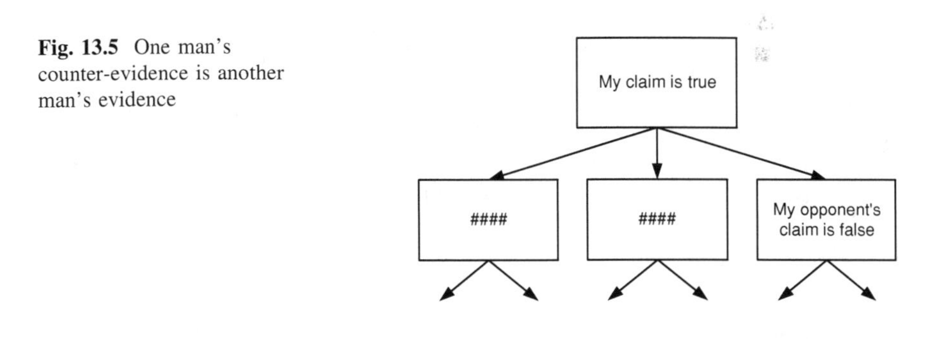

Fig. 13.5 One man's counter-evidence is another man's evidence

That was rather obvious, but the principle can be taken further. Your argument is to persuade someone of something; it will be much more compelling if you can show that, not only do you have Evidence supporting your claims, but also you have Evidence that you looked for, but failed to find, evidence against them.

Going back to the equipment testing example - in such a scenario it is usual to have part of the argument showing that all the tests trace back to the requirements so that you can claim that you have fulfilled those requirements if all tests have passed. Similarly, one would show that all requirements trace to tests, or other means of demonstrating their fulfilment, such as analyses or audits. You can search for counter-evidence to the claim that the equipment does what it should do by additional analyses, user trials and/or testing that is not linked directly to explicit requirements.

For example, you could test what happens to the performance of your computer-based system when you initiate (or simulate) conditions such as communications buffer overflow, slow receivers of output data, enormous log files taking up most of the storage provision, etc. You could also analyse the knock-on effects (if any) of each function not working as intended, and you can extend your user validation trials to see if they can break it with odd key sequences, fast data entry, etc.

This gives you a new claim, "I proactively looked for counter-evidence and failed to find it", or similar. A more compelling argument results, see Fig. 13.6 and compare it with the "There is Something Amiss" example in Fig. 6.10 of Chap. 6.

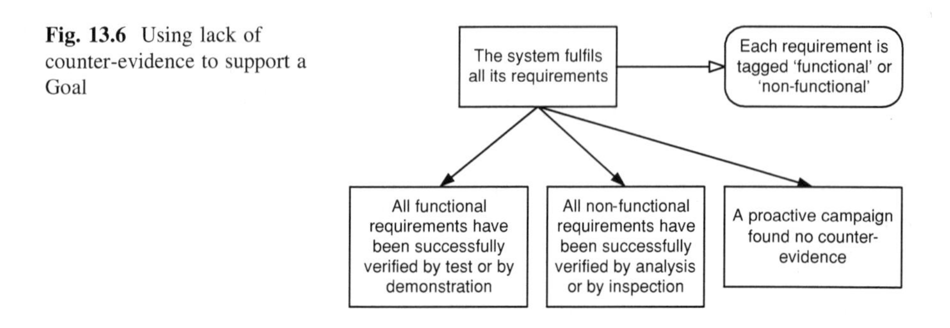

Fig. 13.6 Using lack of counter-evidence to support a Goal

Arguments of the form "All A have property B" are particularly prone to counter-evidence turning up. All swans are white—except for black ones like those on that mill pond down the road. Your search for counter-evidence may well end with you claiming, "Most A have property B" (or restricting the definition of A) but, if it does not, you will have a stronger argument.

13.5 Questions

- Early in this chapter I raised the question, "Can we ever justify it is alright to proceed even if some tests have failed?" I then gave an example showing that we can, but where do you draw the line? In what circumstances would you not proceed; how would you make your case for not proceeding to those in charge?

- Do you agree that it makes a more compelling argument if you show Evidence of attempting, but failing, to find counter-evidence to your claim? If so, do you think that it is something that we should add to generic arguments (Fig. 13.7) whenever possible; is it a practical requirement to impose on pattern authors?

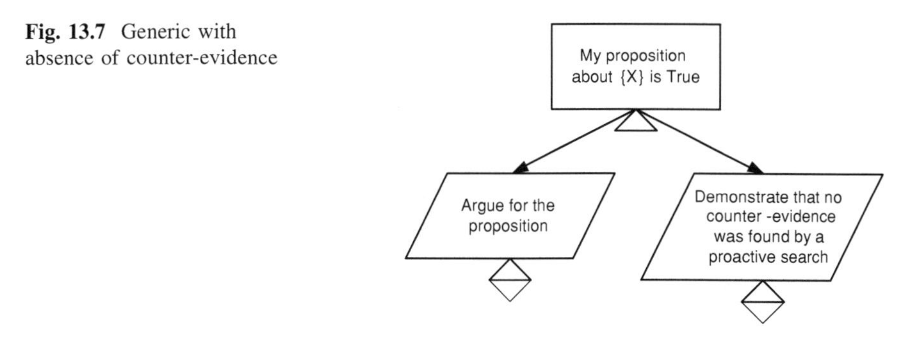

Fig. 13.7 Generic with absence of counter-evidence

- Did you produce a review checklist, as suggested in Chap. 9? If so, add questions about counter-evidence: If any has been found, has it been dealt with appropriately? If you found none, have you used that fact to strengthen your argument?

13.6 Problems

1. Think again about two of the examples from earlier in the chapter. In the first case, the argument about the extinction of an order of fish failed when someone caught an example. In the second, which you may consider a more critical case, an argument for some piece of equipment to go into service did not fail, despite it failing a test. It was acceptable to modify the argument; we did not have to abandon it. Can you see a fundamental difference between the two cases; an indicator that, in one case, it is hopeless to continue, whereas, in the other, it is worth doing a bit more work to see if the argument can be saved?
2. How do you think a nineteenth century palaeontologist with confirmation bias may have attempted to save his claim that the Coelacanth is extinct when he received the report of the remains of one being found in a fisherman's catch? (See Chap. 12 for a definition of confirmation bias)

References

Smith JLB (1956) Old fourlegs the story of the coelacanth. Longmans Green and Co, London

Vaughan D (1996) The Challenger launch decision; risky technology culture and deviance at NASA. The University of Chicago Press, Chicago

Chapter 14
Process Arguments

Abstract Chapter 8 mentioned process arguments when considering sources of evidence. In that context, the process argument is to demonstrate that the right evidence was obtained in the right way. In general, you would deploy a process argument to show that a process is fit for purpose; in particular that it is fit for the purpose for which it is to be, or has been, used. This chapter presents a generic high-level process argument, and discusses decomposition of the key Sub-Goals.

14.1 Process Argument

"Process Argument" is one of those labels that mean different things to different people. In this book, we first encountered the term in Chap. 8, when considering sources of evidence. Our evidence had arisen from carrying out some process and we wanted to argue that it was, in some sense, correct. We wanted to convince the reader that we had obtained the right evidence in the right way.

You may also see a Process Argument presented to support a claim that a process does what is required of it; that it is fit for purpose. For example, I may have a standard process that is always used when installing equipment; for a critical piece of equipment, I would need to argue that the default process is adequate. If it is not, I can use my argument to establish how the process should be augmented to make it fit for this new purpose.

At first glance, these two process argument examples may appear to be arguing the same thing, one is for an evidence gathering process, the other for an equipment installation process, but they are both processes. The difference is that in the first case we are interested in a specific output from the use of a process, whereas in the second case we are looking at the process more in the abstract.

Can you see a similarity to generic arguments, as discussed in Chap. 11, here? Claiming that a process is fit for use in some context is like a making generic claim

J. Spriggs, *GSN—The Goal Structuring Notation*,
DOI: 10.1007/978-1-4471-2312-5_14, © Springer-Verlag London Limited 2012

that applies each time the process is used in that context, whereas claiming that the output obtained on a particular occasion is fit for use corresponds to an instantiation of it.

In reality the correspondence is not exact; the difference is that the 'instantiation' will always require additional argument; it would never be just a case of setting the value of a descriptor in curly brackets. However, I can develop it into a generic argument for instantiation in the normal way, by adding claims for that additional argument to the "fit for use" claim. That generic argument is presented in the next sub-section, overleaf.

Note: If you skipped Chap. 11, I recommend that you go back and read it now, as it introduced generic symbols, which are used in most of the Figures of this chapter.

14.2 A Generic Argument for Correct Process Outputs

Whatever the process is, it will need to be documented; that documentation will form part of the context for the Top Goal. As well as showing that the process is fit for purpose, I also have to show in my argument that someone competent to do so has carried out the process, and that its output(s) have been checked. I have presented the argument as a pattern in Fig. 14.1 below. I used the multiplicity symbol on two Sub-Goals because a process may be carried out by a team, and its outputs may be checked in different ways, such as by inspection or by testing.

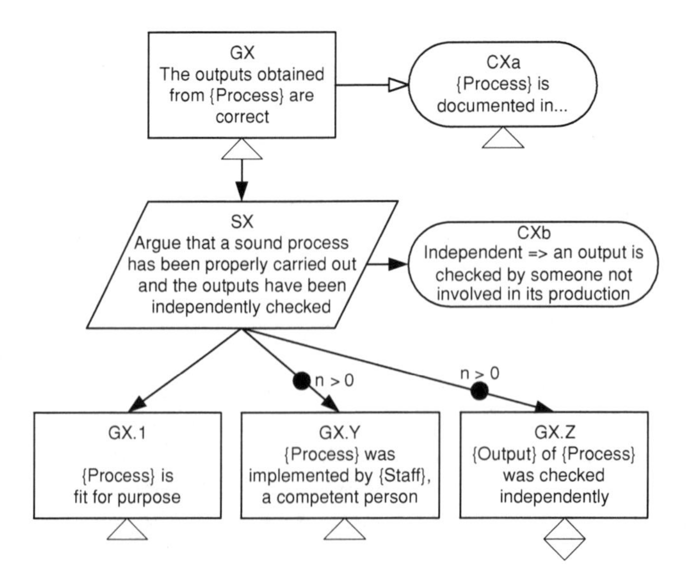

Fig. 14.1 The top level of a basic process argument

As noted in the introduction to this chapter, many authorities regard the process argument as being just the structure below the "fit for purpose" Sub-Goal. They may call Fig. 14.1 a 'product argument', as it is arguing for the correctness of the products of a process.

What you call it does not matter, as long as you are clear what is wanted in each case. Are you arguing for the process itself, for a particular application of the process, or for one of the outputs obtained from a particular deployment of it?

I will not further expand the checking Sub-Goal of Fig. 14.1 but, before getting on to the "fit for purpose" Goal, I will decompose the competency one another few layers to illustrate what some have said is a shortcoming of the GSN.

14.3 {Process} was Implemented by {Staff}, a Competent Person

Note that, in general, "Competent Person" arguments can be quite tricky for (at least) three reasons. The first is that, in the normal run of things, you are not arguing about a process that requires the practitioner to have a specific license or diploma. So what do you use as evidence? They may have had on-the-job training for which there are no explicit training records.

That brings us to the second problem; personnel data are confidential. Your argument may be widely circulated. Even if there were no personal data protection legislation, would you want the identities and skills of your personnel enumerated for all, including recruitment agents and your competitors, to see? You can present your argument in the normal way but, rather than supplying all the evidence, include a statement that the training and competency scheme records, etc., are available for audit by agreement. That way you can control who has access to the data and can set up appropriate confidentiality agreements.

The third reason can be summed up by quoting Juvenal: "Sed quis custodiet ipsos custodes?" (Juvenal 100). This translates as, "But who will guard the guardians themselves?" In our case the equivalent is, "But who will assess the assessors themselves?" Person A is said to be competent at a particular task because Person B has assessed them to be so, but is Person B competent to do the assessment? You may have an accredited competence scheme, but then you need a process argument for it—and are the accreditors competent?

This potential to go on ad infinitum has been held out by some as a shortcoming of the GSN. It is not a shortcoming of the notation; it is a consequence of the clarity that it brings to your arguments. This makes you want to consider things that you may not have thought about in detail before; but what you should be considering is whether more detail is actually required. Is your argument compelling enough without it? The time that you previously spent unproductively worrying about what you had missed out from your argument can now be spent productively in deciding which threads need more detail and which have gone far enough for the job at hand.

If you are providing safety assurance for a critical protection system, you will need to go into lots of detail as to who did what; in contrast, if you are arguing in support of a change of use application for a commercial building, you do not need to address the competence of the draughtsman, for example.

The strategy I used to decompose the competency Goal in Fig. 14.2 was to acknowledge the difference between having knowledge about something and having the skills to actually do it. We also want someone who is competent now to still be so tomorrow, so I also argue that their skills are maintained.

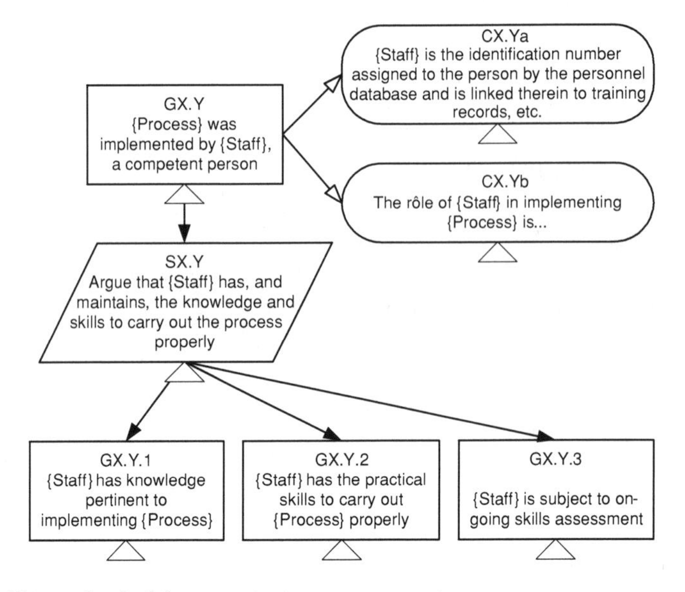

Fig. 14.2 The top level of the competent person argument

Note the Context identifying the rôle that the person in question performs in the process. This is important because it can indicate the amount of detail you need to address at the lower levels of the argument, and upon what aspects you should be focussing. For example, the person who sets up a machine to make widgets will need to exhibit different competences from the person who watches the machine and intervenes if something goes wrong. One must be precise and methodical, whilst the other must formulate and implement effective tactical plans quickly.

14.4 Knowledge

To decompose the leftmost Sub-Goal of Fig. 14.2, I need to identify what the pertinent knowledge is and acknowledge that there are a number of ways for someone to show that they have it to the required level of detail (Fig. 14.3).

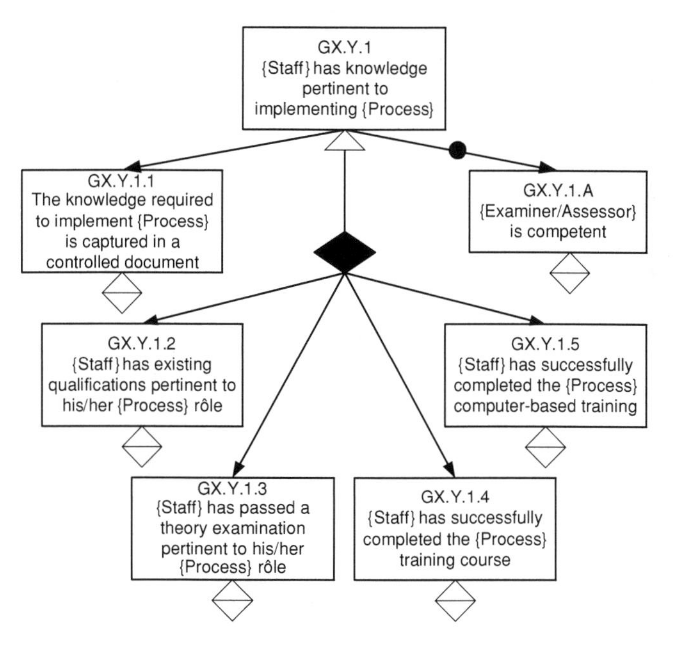

Fig. 14.3 {Staff} has pertinent knowledge

The Juvenal quotation applies to Goal GX.Y.1.A, which also appears in the decomposition of the remaining Sub-Goals at this level. The "Examiner/Assessor" may be a person or an accredited examination board; in decomposing this Sub-Goal, we may also need to argue for the suitability of the examinations, or of the training material used, be it computer-based, human-presented, or guided reading.

14.5 Skills

For the skills Sub-Goal, in Fig. 14.4, I have considered the that there may be an assessment or practical test specifically designed for the rôle to be performed, or that staff may have been subject to a structured interview. This latter option is where you do not let them loose on the process itself, but you check that they would do the appropriate thing in a number of different scenarios. This may done as a prerequisite for entering the practical exam, in which case you would use two of the optional Sub-Goals in your argument; do not use patterns unthinkingly; tailor them to the needs of your specific argument.

For the on-going skills assessment Sub-Goal, in Fig. 14.5, I have assumed that the process is only run every so often, so the practitioners may get "stale"; if it were a continuous process, the on-going skills assessment could be accomplished

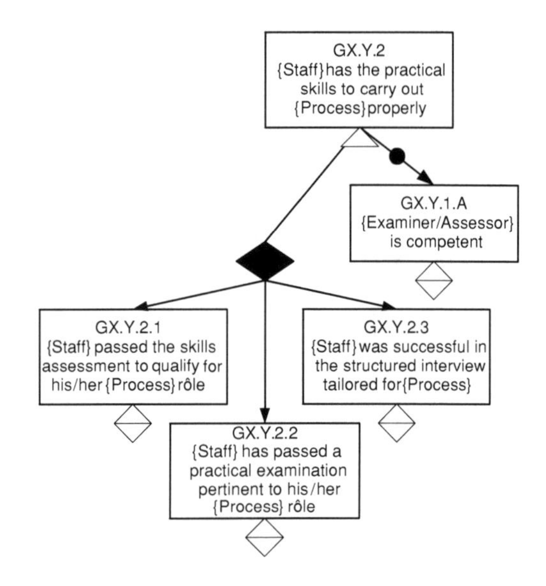

Fig. 14.4 {Staff} has practical skills

by reviewing product quality records to see if there is any correlation between rejected outputs and the member of staff in question.

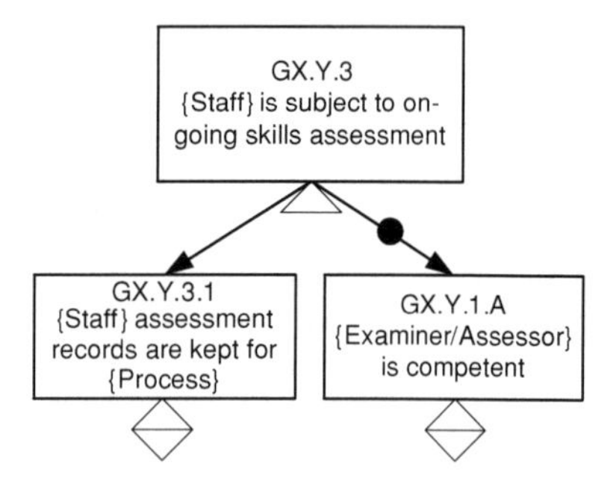

Fig. 14.5 {Staff} maintains practical skills

14.6 {Process} is Fit for Purpose

An obvious prerequisite for satisfying this Sub-Goal is a clearly-stated purpose. Reference this from, or transcribe it into, a Context as shown in Fig. 14.6.

A documented process will have its purpose, or objective, stated on the first page. If not, send it back; get a proper one.

Seriously though, the statement of purpose in the document will not always be the basis of your argument. In many cases, including the example below, the argument will be that the process performs as advertised, but there will be times when you will want to justify that a process is fit for what you used it for, rather than for what it was originally developed.

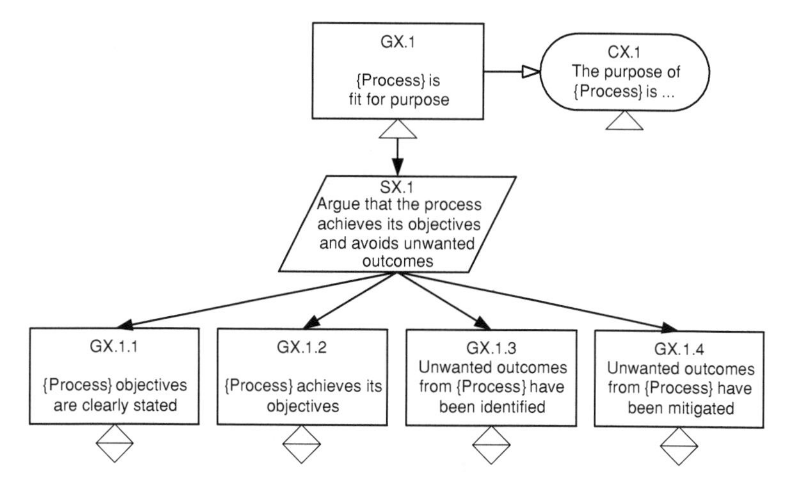

Fig. 14.6 The process is fit for purpose

Imagine the scene; I am off on field trials in a remote spot on the coast with my new multi-sensor array, and I damage the fibreglass radome. Shall I postpone the trial while I go back to base and get a new one? No, not if I can justify using the hull repair process of the nearby sailboat maintenance company. This is a rather contrived example, although I have had to call on the services of local craftspeople when on equipment trials. A more common scenario is where you obtain evidence in support of your argument from processes intended for other purposes. For example, you may be developing part of a safety assurance argument using evidence produced from a logistics support analysis process.

Usually the statement of purpose in a process document just describes intended outcomes but, to be properly effective, the process should also guard against unwanted outcomes.

Unwanted outcomes may be explicitly stated; if they are not, you will have to identify them all yourself and decide how serious it would be if they were to occur. The argument should show that they are unlikely to occur; ideally it will show that they cannot occur. Even if they are stated, I recommend that you identify your own set of unwanted outcomes (preferably before you limit yourself by reading the set given in the document) to ensure that you have covered everything. In any case, what constitutes unwanted outcomes will depend on the purpose of your argument. The hull repair process, for example, has tight controls on the colour and texture of the finished article. I am not bothered by the appearance of the repair, as long as its electrical properties are appropriate and stable; something most yacht owners would not be concerned about.

Note that the pattern of Fig. 14.6 is a straightforward "fit for the purpose for which it was written" argument. You could use the same basic structure for the "fit of the purpose for which I used it" case, but with the Context stating your purpose and the Sub-Goals modified to state your wanted and unwanted outcomes.

This pattern looks reasonable; it covers the points raised in the discussion, so let us try it out on a simple example process.

14.7 Process Argument Example

The process in this example is one of a set used by a fictitious painting and decorating business. The owner manages the projects; he no longer does the work himself, but he wants to maintain the quality of work that underpins his reputation. He has prepared a set of process documents for his subcontractors to follow.

Process 7 is for applying wallpaper to decorate interior walls of houses and offices. Wallpaper requires proper surface preparation before application and there are a number of factors to take into account when measuring a room to establish how much wallpaper (and paste) to buy. Other processes in the set address these preparation topics; there are also processes for related topics, like painting woodwork and replastering ceilings.

The first diagram, Fig. 14.7, is an instantiation of the pattern in Fig. 14.6 for Process 7, "Applying Wallpaper". This is followed by a sequence of diagrams decomposing each Sub-Goal in turn.

After applying a pattern, it is well worth identifying any 'lessons learned' that can be used to improve the pattern, or its supporting material (which I have not supplied here). I will therefore share the lesson of this example, following Fig. 14.11, and modify the pattern accordingly.

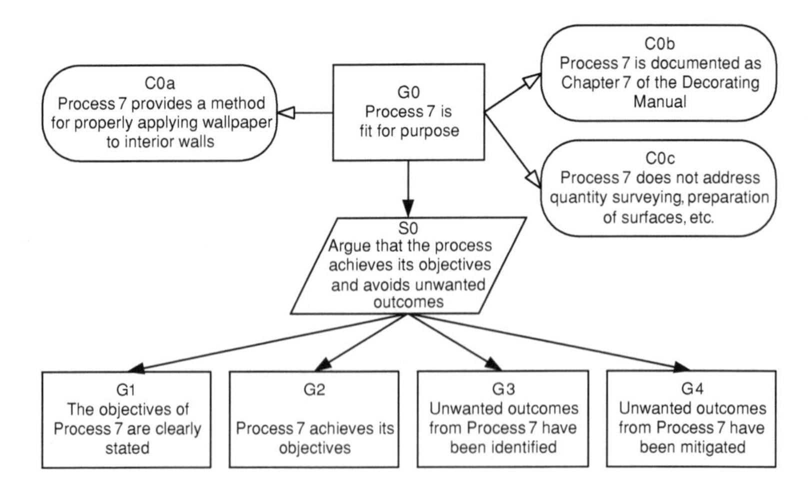

Fig. 14.7 Process 7 is fit for purpose

Note that I have added some Contexts that are not on the pattern; a pattern is like a guideline, it should not be a straitjacket. The Sub-Goals are decomposed in successive figures below; Goal G1 in Fig. 14.8, Goal G2 in Fig. 14.9, Goal G3 in Fig. 14.10 and Goal G4 in Fig. 14.11.

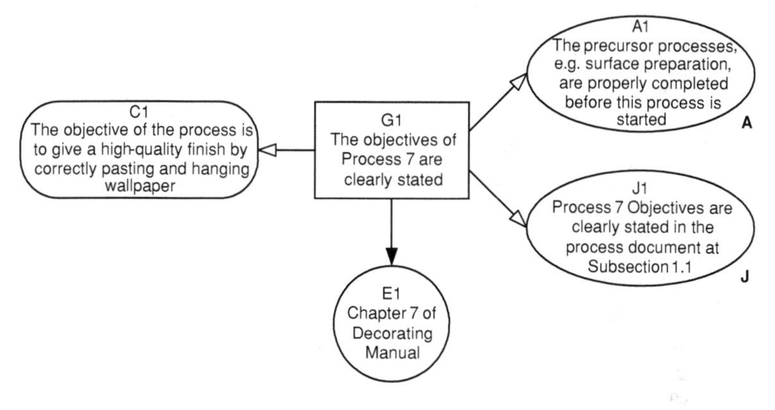

Fig. 14.8 Clearly stated objectives

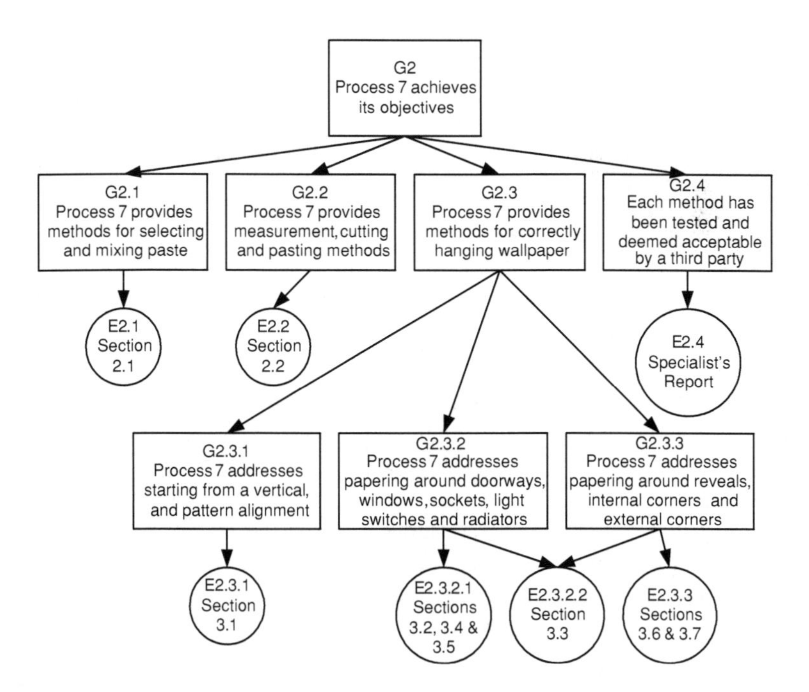

Fig. 14.9 Objectives achieved

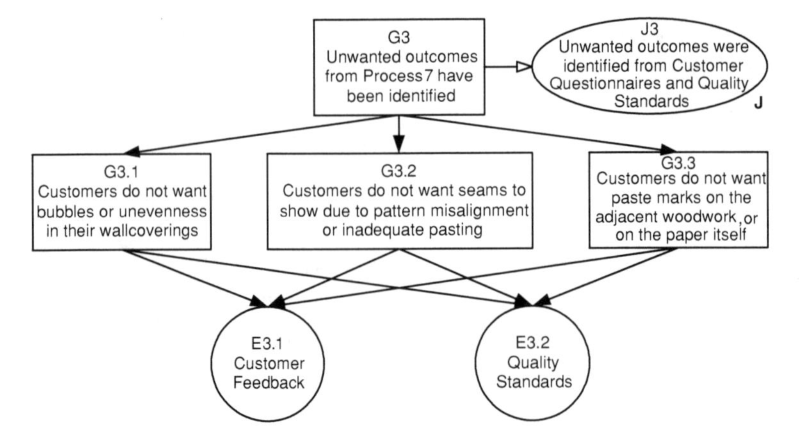

Fig. 14.10 Unwanted outcomes declared

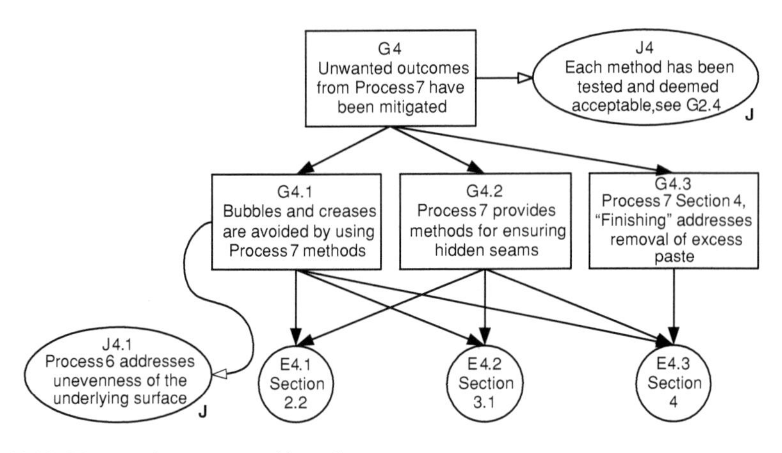

Fig. 14.11 Unwanted outcomes mitigated

Did you notice how a pattern that seemed sensible in discussion was unwieldy to use in practice? The same thing happens with processes themselves; it is not enough to discuss them and write them down, you have to try them out too.

When you first test a pattern, you will find things wrong or ambiguous; fix them and add clarifying notes to the associated documentation. Better still, get someone else to try out your pattern, they will not be so close to it and will find even more ambiguities and improvement opportunities.

So what can we do to improve the pattern? It would be easier to handle the identification of each objective and its achievement at the same time. Similarly, it would be easier to identify and mitigate unwanted outcomes in the same argument thread. See Fig. 14.12 for a proposed new version in which I have taken these points into account.

I wanted to put an Assumption in the revised pattern; it would have provided context for Strategy SX.1.2, saying, "Unwanted outcomes are identified systematically". But this is not an assumption of the argument, it is an assumption I am making about the conditions in which the pattern is used. It is thus a prerequisite of use, and should go in the Documentary Data section of the generic argument (See Chap. 11).

Note the word "risk" in CX.1.2 of Fig. 14.12. We are arguing that we have reduced or, ideally, removed the risk of unwanted outcomes; consequently, this is often called a "risk-based approach".

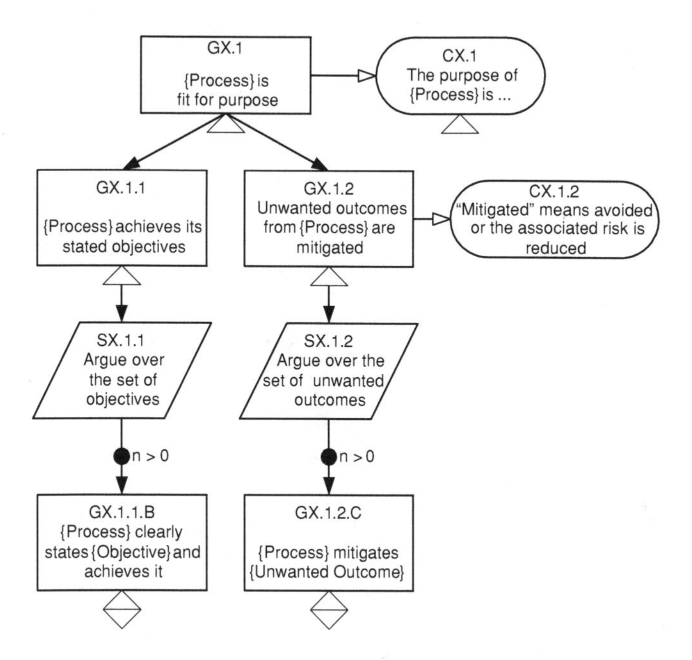

Fig. 14.12 {Process} is fit for purpose pattern

14.8 Other Uses of Process Argument

Up to now I have been using the term 'process' to mean a method to produce some defined output, like a widget, document, or nicely wallpapered room. Process arguments show that the outputs are properly produced whilst avoiding, or at least reducing the impact of, unwanted outcomes. Process arguments can also be applied when the objective is only to avoid unwanted outcomes.

For example, I am installing some large piece of equipment outdoors. I need the subassemblies brought to site by lorry and I need a crane to put them in position. I could just get everyone to turn up first thing on the appointed day and hope for the best, but I do not want any accidents to personnel, or to the sub-assemblies;

I do not want the job to over-run because everyone is in everybody else's way; and I do not want to present a hazard to the public. I avoid these unwanted outcomes by using scheduled sequences of work, banksmen and physical barriers.

I can use a process argument to demonstrate that the resulting "choreography" of lorries, assemblies, cranes, personnel, etc., is adequate; that the risk has been reduced enough.

I can also use the process argument pattern to argue that my tools are fit for purpose. Figure 8.7 of Chap. 8 (or see the answer to Problem 3 thereof, which has another instance), has a Sub-Goal that is supported by a "Tool Verification Report". Such a report could contain a tool argument based upon the process argument pattern; the whole one, with the Sub-Goals addressing competent people and checked results. A tool (or process) argument does not have to be relegated to an external document, there is nothing to stop you just copying it into your overall argument; the Top Goal of the pattern becomes one of your Sub-Goals (with suitable Context and/or Justification, of course).

A pitfall exists for tools that does not arise with processes. It is not unusual to see tools advertised that have been "certified". This means that they comply with some standard, which will be identified on the certificate; it does not mean that the tool is fit for purpose. Having a certificate for your tool may make your argument easier (and it may not); it does not replace it.

I started this chapter by referring back to Chap. 8, where an evidence gathering process was required. An example of such a process is one for in-service failure reporting. If I want to justify the use of a particular piece of equipment on the basis that an equivalent item is already in operation, I would need to show, inter alia, that it has been sufficiently reliable for the new application. This topic was touched on in Problem 1 of Chap. 12. The argument would be based on failure reports; but how does the reader know that I am using all the reports collected or, for that matter, that a report is generated for all failure events? For example, it is possible for a piece of equipment to appear reliable, when it is actually down to highly-skilled operators, who have developed works-around for each type of failure. I need to construct the process argument such as to demonstrate that reports are collected on every failure and that the analyses consider every report. I could produce a generic process that is instantiated for each equipment I have in operation. I can then base the generic process argument (and its instantiations) upon the structure of the process, or vice versa.

14.9 Questions

- Do you think that it would be practical to construct a process argument at the same time as you are designing the process? If you were to do this, would it result in a 'better' process than if you had not prepared the argument? Could you construct the argument first, and use it to specify the process, in a similar manner to the checklist generation of Chap. 5? If either of these approaches would produce better results, what is stopping us from doing things this way?

- You may be constructing a stand-alone process argument, for example, using it to help develop a good process, but often you will be using a process argument in support of a larger argument. For example, in a system safety argument, you may be arguing for some property of the system, bringing evidence in support. You will probably need a process argument to show that it is the right evidence. You may also bring process arguments in support of a claim that what you have built has the same properties as intended by the design. What is the best way of linking the process argument to the whole? Can you remember how we linked arguments in earlier chapters? The next chapter summarizes an alternative method.

14.10 Problems

1. Use the revised pattern of Fig. 14.12 to re-do the "Applying Wallpaper" process argument.
2. Modify the pattern in Fig. 14.1 to argue for the outputs of a computer-based tool. Replace {Process} with {Tool} in the Top Goal, then change the other elements as necessary.

Reference

Decimus Iunius Iuvenalis (anglicized to Juvenal): Satire VI, ~ 100 CE

Chapter 15
A Brief on Modular GSN

Abstract Dr Tim Kelly formalised the Goal Structuring Notation for safety assurance and, in 1998, published it in his University of York doctoral thesis. Since then, development work has continued at York, and elsewhere. This chapter briefly summarises one of those developments, the concept of modular arguments; my intention is to give you enough to decide whether you want to find out more. No problems are set.

15.1 Links to Other Arguments

In one of the questions of the previous chapter, I asked you to recall how we had linked arguments earlier in the book. This was in the context of process arguments brought to justify evidence, or in support of a claim. Figure 15.1 shows an example in which an external process argument is used to justify a test method. In this example I have not explicitly referenced the documents so, if it were real, I would have supplied references in a table below the diagram.

In Fig. 15.2, I have instantiated a pattern from Chap. 14, with a rather simplistic appeal to Evidence for all the Sub-Goals other than that requiring a process argument. That argument is supplied externally; in effect it 'plugs in' to the Goal Structure, in contrast to the previous example, where it was contextual.

In both of the preceding examples we made use of someone else's argument, or one that we prepared earlier. In the first, a Justification was used to link to that externally supplied argument, in the second our Sub-Goal was a copy of the Top Goal of the external argument. I can re-draw the first example such as to state the external Top Goal in a similar way, see Fig. 15.3. The, possibly unexpected, ramification of this is that a Goal can appear as Context, but it is an external Goal, what Dr Kelly calls an "Away Goal" (Kelly 2003).

J. Spriggs, *GSN—The Goal Structuring Notation*,

DOI: 10.1007/978-1-4471-2312-5_15,

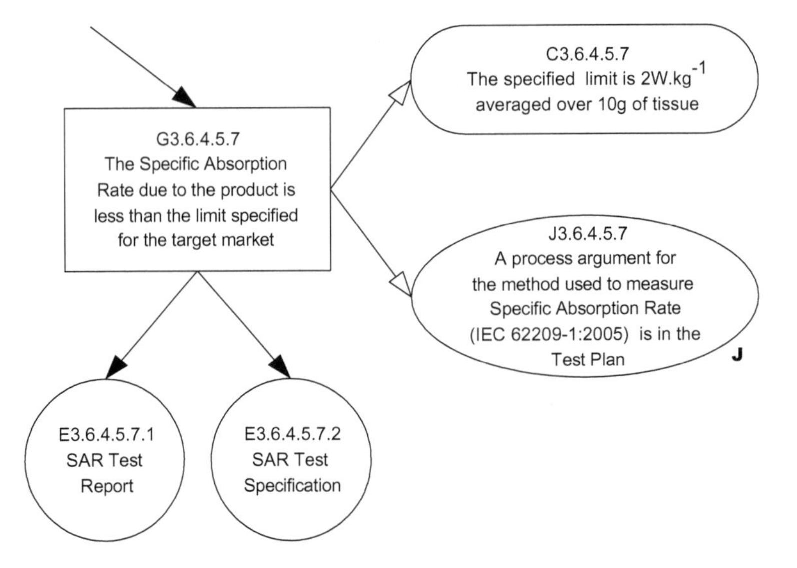

Fig. 15.1 Justifying Evidence with a process argument

15.2 Away Goals

Apart from appearing as contextual information, an Away Goal is just like a normal Goal in most respects. It contains a proposition that can be either true or false; we claim that it is true. Its distinguishing feature is that it is not decomposed locally; that is done elsewhere, i.e. away. This means that, although Thread of Argument arrows can come in the top, none leave at the bottom. Instead we have a named ‘socket’ into which the external argument ‘plugs’, see Fig. 15.4.

It is not really a socket; the grey shape represents the package of argument in which the Goal is decomposed; it contains the argument that supports the Goal. In the example, the package is called TestArgs, short for Test Arguments (and very short for “Process arguments that underpin the test methods used”).

Note also that the Goal identifier is a name (again, horribly foreshortened); it is not a G-number. This convention is to avoid having a plethora of Goals G0, G1, etc. from external sources. There is not a great deal of room for a Goal identifier, so it is a short string of characters, and so may still be prone to duplication. In a new collaborative project, where your colleagues and suppliers are preparing bits of argument for you, I recommend agreeing a naming convention early on (and avoid the temptation of including the name or number of the project as a prefix; there is not much room in those symbols).

You cannot do much about the names of Goals you have bought-in, or are re-using from other projects. I recommend that you just live with them; do not attempt a mapping between your nice logical naming convention and the random

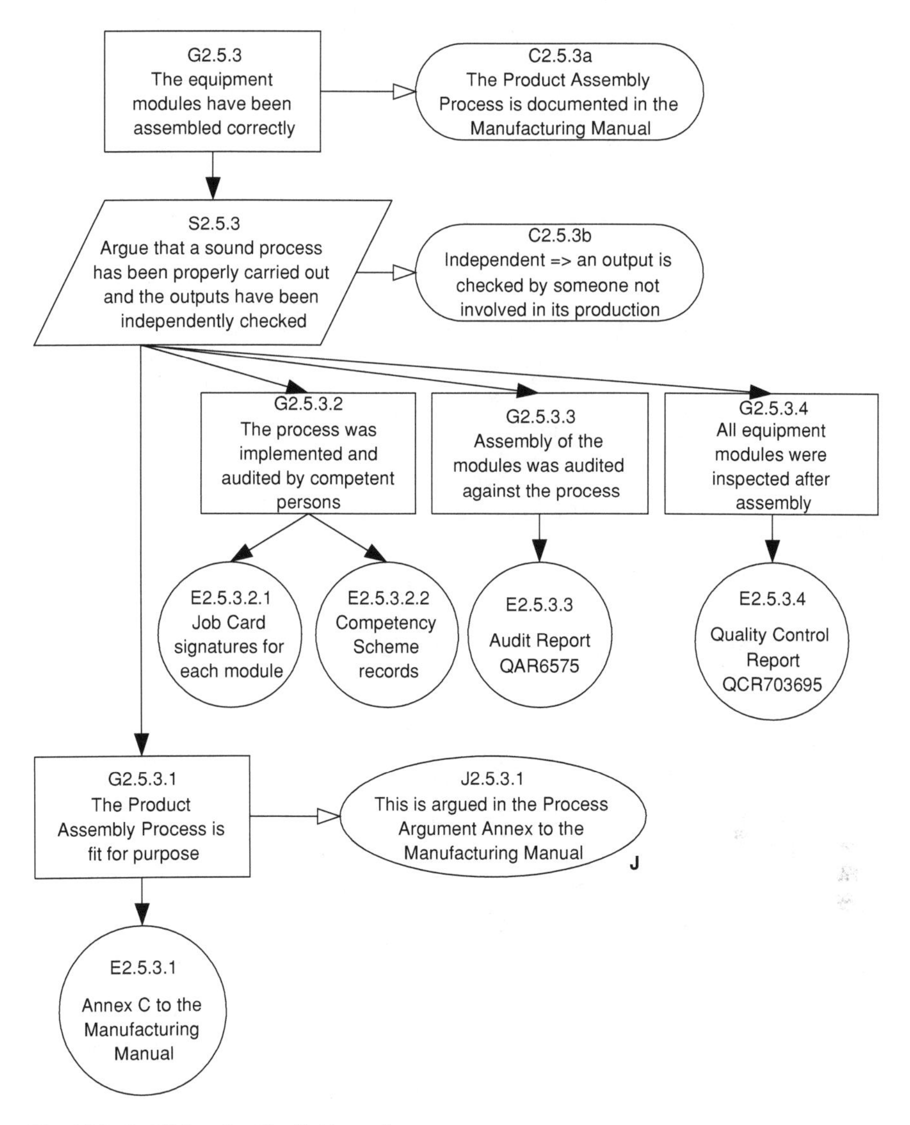

Fig. 15.2 Justifying that the Evidence is a process argument

strings of the extant Goals. Mapping just adds another layer of complexity and, hence, confusion; rarely does it clarify anything.

I introduce the 'Goal as contextual information' example (Fig. 15.5) first as, although familiar with Justification referring out to someone else's argument, I must admit that I had not thought of it as an external Goal until I saw Tim Kelly's presentation (Kelly 2003).

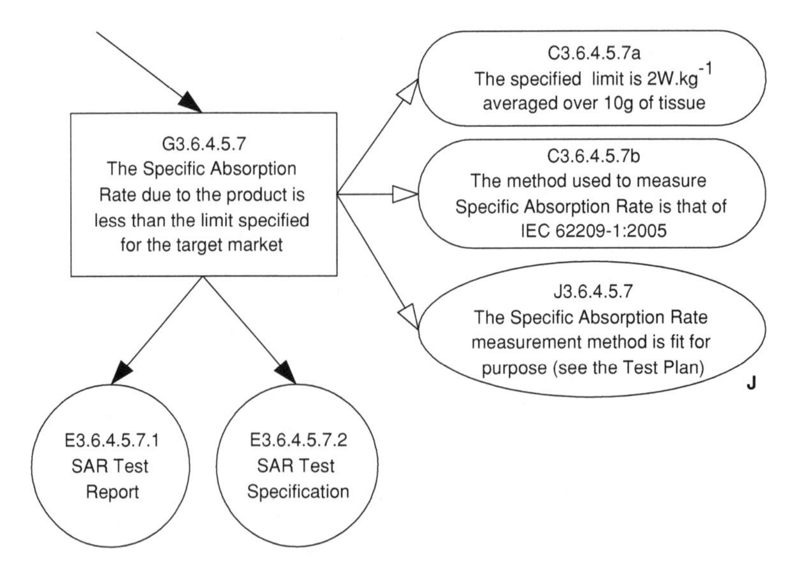

Fig. 15.3 Justifying Evidence with an explicit process claim

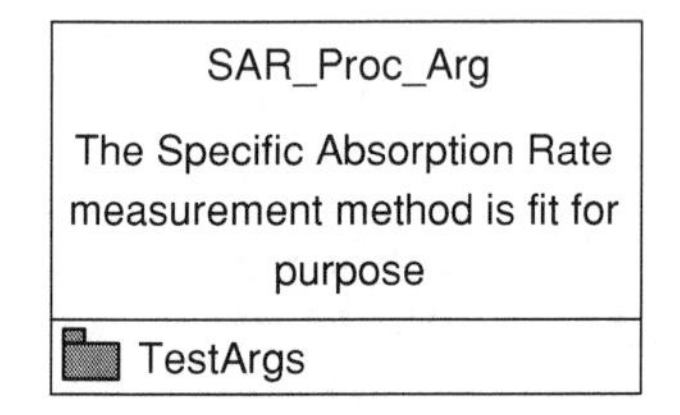

Fig. 15.4 Modular GSN symbol for an Away Goal

Of course, an Away Goal can also appear in a Goal Structure just like any other Goal (but with no outgoing Thread of Argument arrows). The Away Goal is similar in principle to the (deprecated) Goal Developed Elsewhere symbol of Chap. 8, but it provides the essential external reference missing from that symbol.

Figure 15.6 shows an example with Away Goals supporting a Goal directly via Thread of Argument arrows. In a 'real' argument I would probably have used a Strategy to explain the decomposition (and also I would have made claims about the measurement equipment used and its calibration).

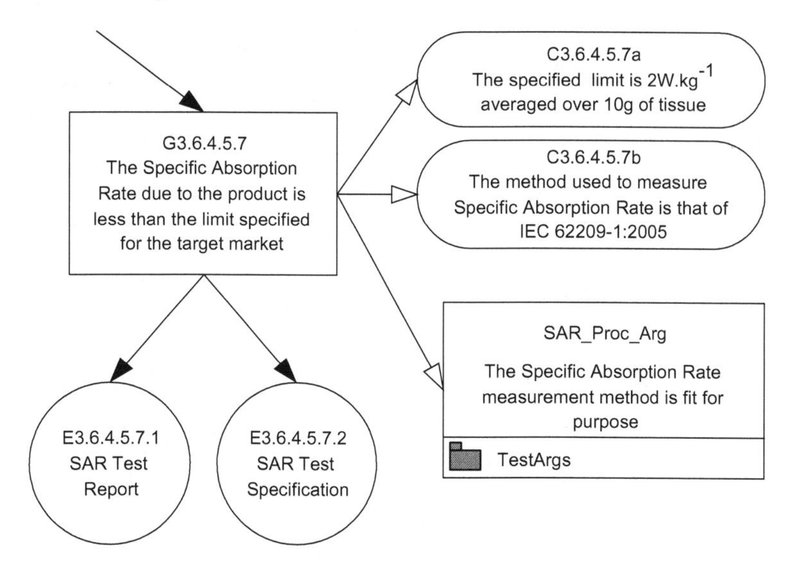

Fig. 15.5 Justifying Evidence with an Away Goal process claim

15.3 Modules and Modular Argument

The ‘package of argument’ that I referred to earlier is more properly called a Module, hence “Modular GSN”. It was appropriate to refer to it as a package, as the symbol was derived from the package notation used by many system and software engineers. In the Unified Modeling Language (UML 2005) the package is a construct used to group low-level elements into higher-level ones. Each element within such a package must have a unique name and some of the elements may be ‘private’, i.e. not visible from outside. An Away Goal, in contrast, must be visible from outside, even if its decomposition is not; it must be ‘public’.

The Module symbol is the shape of the grey icon on the Away Goal symbol. This is apparently meant to look like a tabbed folder, see Fig. 15.7. The symbol contains an Identifier, e.g. TestArgs, and a Description, e.g. “Process arguments that underpin the test methods used”. However, the convention is to aim for more of a newspaper headline style; just nouns and qualifiers, no verbs.

But why do we need a symbol for a Module, if we explicitly invoke the Away Goals in our arguments? Having a symbol for a Module enables us to build and present modular arguments, which may be used, for example, in agreeing the work shares when splitting production of an assurance argument (Fig. 15.8). This is only a rough structure; it does not identify the Away Goals that are being invoked.

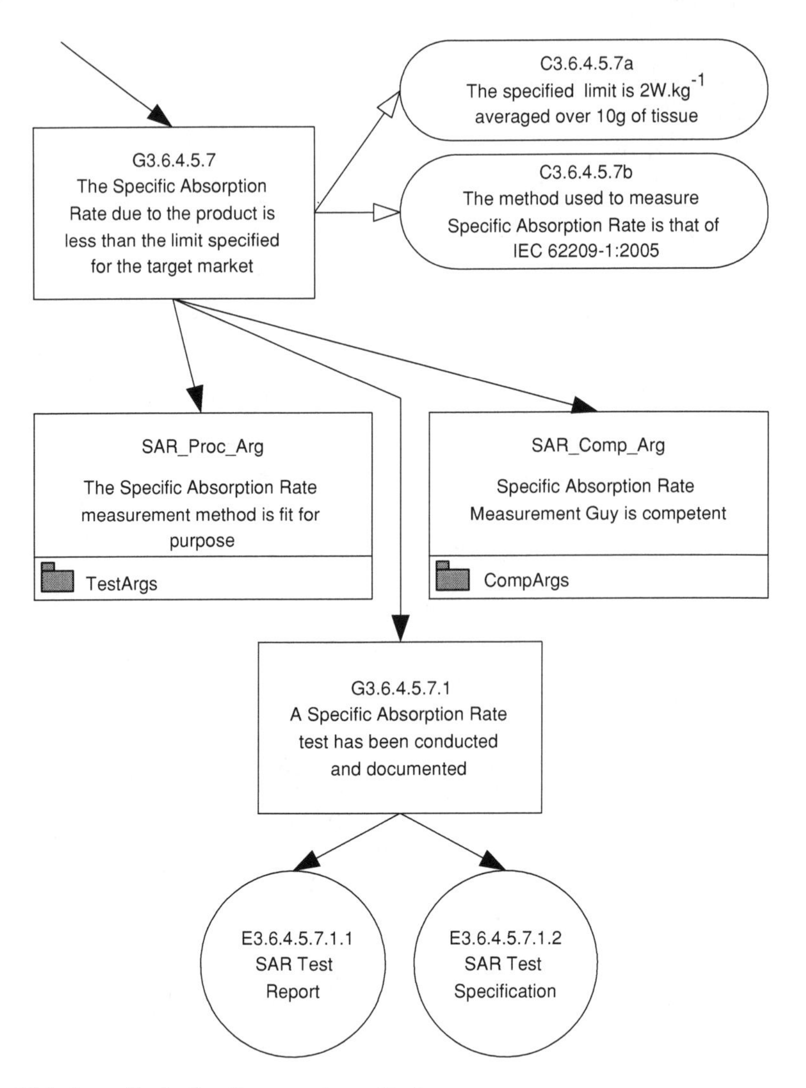

Fig. 15.6 Away Goals directly supporting a Goal

Compare Fig. 15.8 with the top level of a basic process argument as shown in Fig. 14.1 The arguments that all the various types of test data are correct and have been checked is contained within the TestDataArgs Module, these are supported by process arguments for the various test methods (TestArgs) and competence arguments (CompArgs). I invoked Away Goals from these Modules in the Specific Absorption Rate example in Fig. 15.6.

Some readers will have thought me unnecessarily informal in referring to "Specific Absorption Rate Measurement Guy" in the competency Goal of Fig. 15.6, but I was just being intentionally ambiguous. In that example, I was

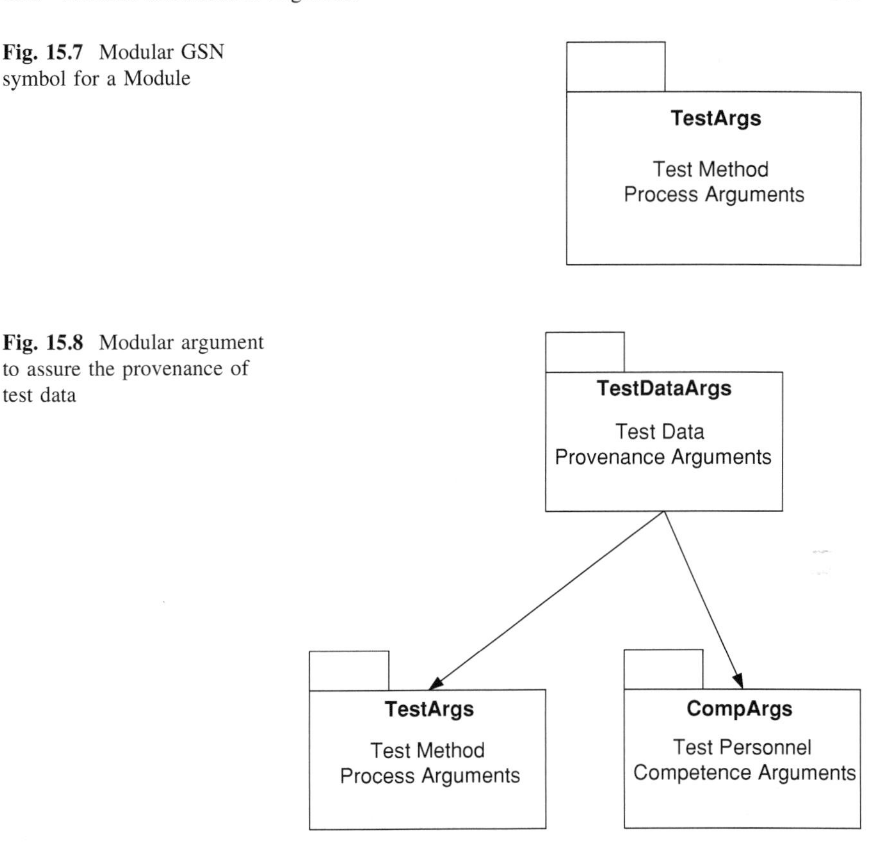

Fig. 15.7 Modular GSN symbol for a Module

Fig. 15.8 Modular argument to assure the provenance of test data

suggesting it was a particular person, but in general it could be an anonymous rôle holder. I could argue for the rôle and the group of qualified people that I have available to fulfil it. See how the CompArgs Module neatly avoids the confidentiality problem highlighted in Chap. 14. The 'public' Goals claim competence of anonymous rôle holders, whilst the evidence specific to each of the cohort of qualified rôle holders remains 'private'. A Regulator, or other trusted third party, could review and agree that the content of the Module is appropriate, other parties would just see a 'certificated interface'.

Within certain constraints, a modular argument may also be used to assure a modular system. Indeed, one of the stimuli for developing Modular GSN was the desire to have modular safety cases for modular avionics, i.e. an integrated system in an aircraft made up of standardised Line Replacement Modules in cabinet assemblies, rather than the traditional ad hoc arrangement of Line Replaceable Units (IAWG 2007).

This all works well in principle, and modular arguments are a good way of building, maintaining and controlling a portfolio of assurance argumentation to go with your portfolio of products, but I can imagine it becoming a bit of a nightmare in a complex system with many suppliers and stakeholders. A wide variety of Modules will come from a wide variety of commercial and other organisations. "My argument depends on your argument, which is hidden from me, so you are liable if anything goes wrong." "No I am not, because you have invalidated my argument with your implicit flow down of Context and Assumption." Clearly more is required than just being able to see someone else's Goal; before looking at that, there are a few more symbols to introduce.

15.4 Other 'Away' Elements

Consider again TestArgs, which is the Module used as an example for the GSN symbol in Fig. 15.7. It is a bag of process arguments assuring our test methods. You may be doing a test safety argument and make a claim about the behaviour of the item under test whilst Specific Absorption Rate, for example, is being measured. The test method is thus contextual information; you could get it from the TestArgs Module by referencing it as Away Context, see Fig. 15.9. It would, of course, have to be made 'public' in the Module for you to do this.

Fig. 15.9 Modular GSN symbol for an Away Context

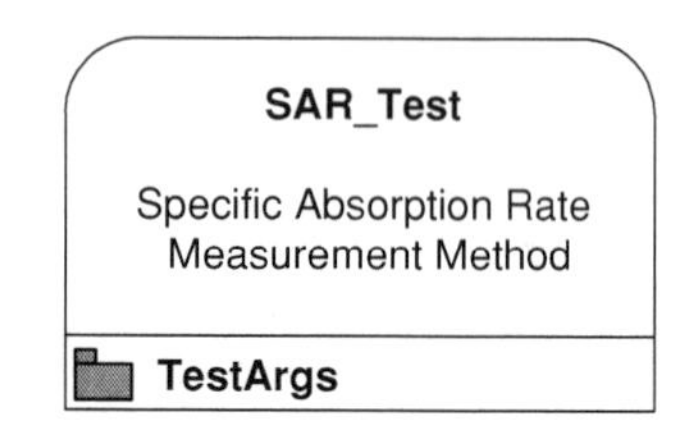

You could also conceive of a need to reference Evidence from a Module. For example, one of our test methods uses an instrument the output of which needs to be corrected with a polynomial. You may wish to show that the instrument is suitable for some other use and invoke the correction coefficients as 'Away Evidence' from the TestArgs Module, which writers on modular argument tend to call an Away Solution, see Fig. 15.10. Again, this needs to be 'public' in the Module.

Fig. 15.10 Modular GSN symbol for an Away Solution

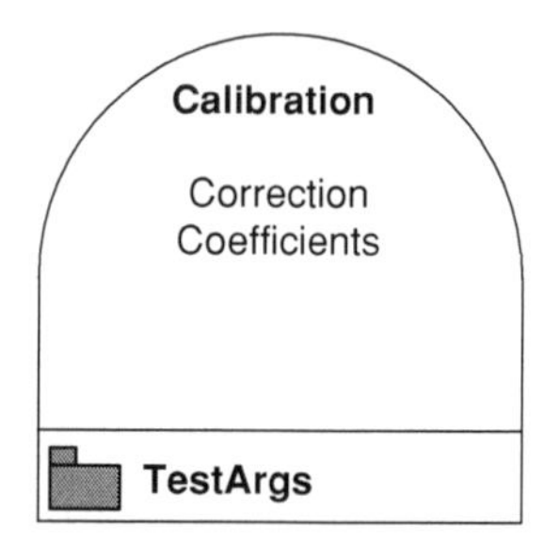

What else; can you have an Away Assumption? Well, no; Assumptions tend to be more personal, you may make them about the remote Module, but you will not want to re-use someone else's Assumption from that Module. Away Justification is already covered. If a Justification is giving simple rationale, or highlighting a fact, it will be local. If it is providing a supporting argument, implement it as an Away Goal; there is no need for an additional symbol. An Away Model would be viable but, as the Model is deprecated, use an Away Context instead.

There are other new symbols that you would use inside a Module to designate an element as 'public'. Figure 15.11 shows the symbol for a Public Goal corresponding to the Away Goal of Fig. 15.4. You can construct the 'public' symbols for other 'away' elements in a similar manner.

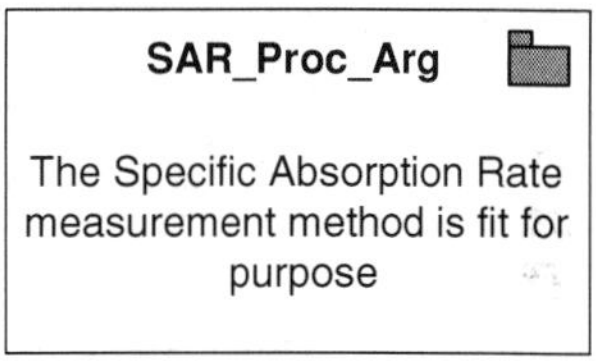

Fig. 15.11 Modular GSN symbol for a Public Goal

SAR_Proc_Arg

The Specific Absorption Rate measurement method is fit for purpose

Some means of managing these interfaces is required, both to track completeness during development and to assess impact of changes during maintenance. A cross-reference table may be sufficient; it should list which Modules require support and which provide it, and explicitly identify which 'public' elements resolve the 'away' elements.

15.5 Interface Problems

You cannot (although I am sure that there are those who will) just outsource the difficult bit of your argument to an Away Goal in the hope that someone will support it. Presumably you can imagine building your argument by selecting Away Goals, as appropriate, from those in a catalogue. What about those who wrote the arguments for you to have away; what did they do?

If we have constructed a Module, how can others use it? We need to define a public interface; we need to rigorously define the connections and constraints in much the same way as we had to provide documentary data to go with an argument pattern in Chap. 11. Other users will need to be able to see the Away Goals we support; we will declare them as the public Goals of the Module. If we also supply public Contexts and/or Solutions, they will need to be declared too. There will be times when we need to invoke Away Goals, Away Solutions and/or Away Contexts from our Module; we need to identify those too, so that the users will know that their problems are not completely solved by this Module.

Going back to the plug and socket analogy, our Module provides plugs at the top that others can use as Away Goals or Away Solutions; there are also plugs at the side that others can use as Away Contexts. The Module has sockets on the bottom so that we can take advantage of others' public Goals and Solutions; also there are sockets at the side for getting to others' public Contexts or Goals.

Do we need anything else? Yes, as implied earlier, we will also need visibility of contextual information across the interface. I suggested that it may be possible to flow contextual information into a Module that invalidates it; the reverse also holds, internal contextual information may restrict support of a public Goal far more than the remote user expects.

For example, I invoke an Away Goal, "The Protection System is available for use when required", from someone else's Module. I want that system to be continuously available, but the Module contains an Assumption that its use will be demanded no more than once a week, and that preventative maintenance is carried out after each event, making it unavailable for an hour. I need to know about that Assumption! Just having an 'away' to 'public' cross-reference table, as suggested in the last section, is not sufficient if you are using elements from someone else's Modules.

Could we make Module authors develop their public Goals context-free? No, that would strip the GSN representation of most of its power.

There are solutions to these problems, for example by the use of a so-called Contract Modules that, in effect, argue the interface (and introduce some new symbols to do so). If you wish to pursue that topic further, this is where we part company. This book is intended to be an introductory guide to the use of what I have called "Core GSN" in presenting arguments; I have provided this overview of Modular GSN to let you know that it is available. If Modular GSN looks appropriate for your application, I recommend that you go and look at the Industrial Avionics Working Group documents (IAWG 2007), which are available on the Internet; just search for "IAWG-AJT-301".

15.6 Questions

- This chapter has presented a brief (and incomplete) overview of Modular GSN; do you think it too complex a concept for your applications of GSN? Maybe it is, but do not be put off by this; try some simple examples where you have everything under control. Can you partition your sort of argument into re-usable modules? If so, can you use Modular GSN to provide a controlled means of building a library of argument segments?
- Have you produced an author's or reviewer's checklist, as per Chap. 9? If so, add questions to check the use of the symbols you will encounter in modular arguments (and the clear definition of module interfaces).

References

Industrial Avionics Working Group (2007) Modular software safety case process; published in two parts, A—process definition, B—guidance, IAWG-AJT-301, Issue 2, 2007

Kelly TP (2003) Managing complex safety cases. In: Redmill F, Anderson T (eds) Current issues in safety-critical systems. Springer-Verlag

International Standard ISO/IEC 19501 (2005) Information technology—Open Distributed Processing—Unified Modeling Language (UML)—version 1.4.2, International Standards Organization and International Electrotechnical Commission, 2005 (Note that this is not the version of UML used by most practitioners, but it is the most recent formally standardized one.)

Chapter 16
A Summary of Goal Structuring Notation

Abstract This chapter presents a Revision Guide, a summary of the Goal Structuring Notation. It illustrates each symbol in turn with the rules for its shape and content, plus references to examples elsewhere in the book. Note that it does not cover 'Modular GSN', see Chap. 15 for that.

16.1 Purpose and Scope

The purpose of this chapter is to provide a Revision Guide on the Goal Structuring Notation, GSN.

It covers only those symbols regarded as core GSN by many practitioners and it excludes extensions, deprecated symbols, notes and labels. For each symbol, I give the rules for shape and content, plus a reference to the chapter of the book in which I first introduced it, which will tell you how to use it.

16.2 Thread of Argument Symbols

16.2.1 Preamble

When you use GSN to represent an argument, the Goals are the claims; the Top Goal is the proposition for which you are arguing. Flow of argument is essentially down a goal structure; one Goal to many, and on to many from each of those. Use Strategy symbols to explain the decomposition of a Goal into Sub-Goals, Chap. 6 gives a process for doing this. Terminate the threads of argument with Evidence symbols, sometimes called Solutions, which support the lowest-level Sub-Goals.

J. Spriggs, *GSN - The Goal Structuring Notation*,
DOI: 10.1007/978-1-4471-2312-5_16,

16.2.2 Goal

The fundamental entity in the Goal Structuring Notation is the Goal, introduced in Chap. 2. It represents a claim, the truth of which we are going to demonstrate by argument, by decomposing it into sub-claims (represented by Sub-Goals).

The GSN symbol for a Goal is a rectangle (Fig. 16.1); it encloses the text of the Goal Statement, i.e. the claim. The Goal statement should be of the form subject-verb-object and as succinct as you can make it without using abbreviations, or missing out words. The statement can be either true or false; we argue that it is true.

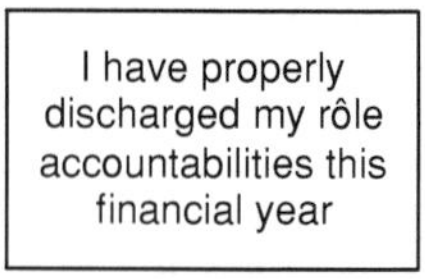

Fig. 16.1 Goal

I have properly discharged my rôle accountabilities this financial year

Note that the Top Goal, and each Sub-Goal, is just a Goal; the difference in name just identifies its relationship to other Goals.

For examples of both valid and invalid Goal statements, see the Problems of Chap. 2; in the Answers section I give examples of how to make a claim more succinct, and also suggest how you may change the way you present your claim depending on the purpose of the argument.

16.2.3 Thread of Argument Relationship

Connect Goals to their Sub-Goals and to other argument elements using solid-headed arrows; these represent the Thread of Argument Relationship, see Fig. 16.2. The arrow links the bottom of a Goal (or other argument element) to the top of another argument element, into which it points.

Fig. 16.2 Thread of argument relationship

You can read this arrow in Goal Structures as "is supported by". It links a Goal, representing a claim, to another argument element that supports the claim. I discuss how you can construct arguments from Goals in Chap. 4.

Note that these arrows do not have to be straight lines; I have made them so in most of the examples in this book just because it is easier so to do in the drawing tool I used. Threads of Argument can cross over each other if necessary; however, for clarity, it is best to avoid crossovers if at all possible.

It is a valid construct to have a Thread of Argument fork to two (or more) sub-elements, but I recommend that you avoid this for clarity and either use two (or more) separate arrows, or use a Strategy to explain the split (See Chap. 11 for a brief description of where you may see such a fork).

16.2.4 Strategy

I introduce Strategy in Chap. 6; use it to give a rationale for your choice of Sub-Goals. It explains how the Sub-Goals address the parent Goal.

The GSN symbol for Strategy is a parallelogram, see Fig. 16.3; it encloses the statement of Strategy, i.e. the rationale. Strategy is typically phrased using "Argument" or "Argue" as a prefix, such as "Argument over each diagnostic criterion in turn", or "Argue from construction".

Fig. 16.3 Strategy

Argue in turn for each
group of accountabilities
defined for my rôle

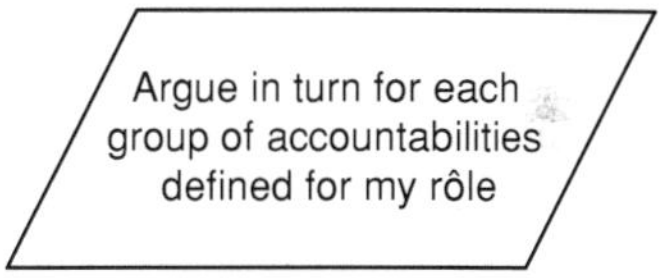

It is also valid just to express your strategy as an imperative statement, for example, "Show that a sound procedure has been implemented, by competent personnel, to select appropriate crew for the mission".

You can use Strategy for emphasis as well as for explanation. In this case you may wish to identify the type of argument used with your Strategy Statement, for example, "Argument by analogy", or "Independence Argument". I recommend that you always use Strategy for clarity in Generic Arguments, or patterns; see Chaps. 11 and 14 for examples of this usage.

More than one Strategy may be deployed below a single Goal when you wish to emphasise a partition of your argument, for example to present it from different viewpoints, or to split the argument for a claim of the form "A and B are true" into separate sub-arguments for A and B. I give examples with multiple Strategies in Chap. 6.

Thread of Argument arrows link the bottom of a Strategy symbol to the tops of the Sub-Goals that implement the strategy, the Sub-Goals that support the parent Goal. Link those Sub-Goals in a similar manner to other Goal or Strategy symbols to form the argument structure.

16.2.5 Evidence

There comes a point when further decomposition of a Sub-Goal is unnecessary, because you can bring a piece of evidence to demonstrate the truth of this lowest-level claim. For example, I may claim that "the candidate has had appropriate training for the job" and demonstrate that this is true by presenting the candidate's training certificate. I would have to have claimed elsewhere that the training was indeed appropriate, and could have supported that with a copy of the syllabus.

Represent such items in the Goal Structure with the Evidence symbol that I introduced in Chap. 8. Evidence is shown as a circle with a (qualified) noun in it (Fig. 16.4); this may be the name of the evidence item, or some other means of referencing it. Use the Thread of Argument arrow to link to the Evidence from the Goal that it supports.

Fig. 16.4 Evidence

Chapter 8 discusses how to make it clear that the Evidence does indeed support the Goal, whilst Chap. 12 addresses what to do when you do not have all the evidence items you expected to have when you constructed your Goal Structure. Chapter 13 looks at what we can do in the, potentially more damaging, situation where you find plenty of evidence, but it tends to indicate that your claim is false.

Note that some authorities refer to the Thread of Argument arrow as the "is solved by" relationship, rather than "is supported by" and, consequently, refer to the Evidence symbol as a Solution.

16.2.6 There is More to Come

In Chap. 9, I recommend that, if your argument is large or complex, you get it reviewed by someone else early on in its development. When you do this, you will want to indicate which bits are unfinished, which Sub-Goals have yet to be decomposed. GSN has the "Goal to be Developed" symbol for this purpose, see Fig. 16.5; some call it the "Undeveloped Goal" symbol.

Fig. 16.5 Goal to be Developed

The syntax, meaning and labelling is just the same as for any other Goal. The addition of the diamond just means that you do not present further decomposition, for now at least, and so no Thread of Argument arrows come from this Goal.

You do not have to stop decomposition at a Goal; it is sometimes desirable to indicate what will happen next by presenting a "Strategy to be Developed", as shown in Fig. 16.6 below. This way you can get feedback from your reviewers on your intended next steps, as well as on what you have done.

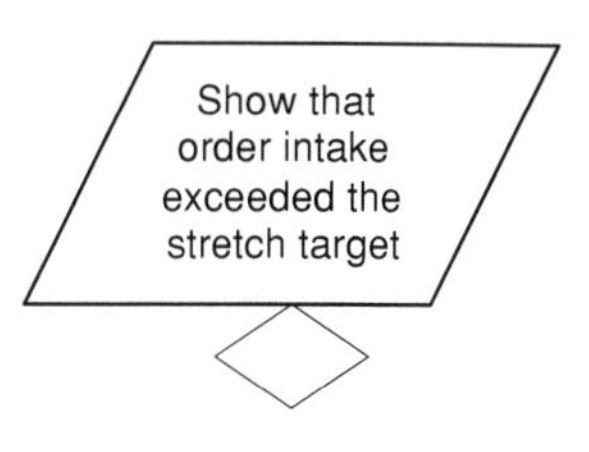

Fig. 16.6 Strategy to be Developed

Using this symbol in a review copy of your argument does not commit you to having a Strategy there in the final version. Decomposition may turn out to be straightforward, so you can replace it with a few obvious Sub-Goals.

Note that, in a small argument, it is usually obvious which threads you have completed and which remain for you to address. These "to be Developed" symbols are most useful in a complex argument that is presented over many pages. In such an argument, you will decompose (usually on another page) a Goal shown with no Thread of Argument arrows coming out of it, whereas the diamond tells the reader not to expect any such further decomposition.

16.3 Contextual Information Symbols

16.3.1 Preamble

As I noted in Chap. 1, there are other notations available for presenting arguments. What makes GSN superior to the others is explicit contextual information.

Context in a Goal Structure allows you to state your claims succinctly without the use of the supporting text required by other notations. Context allows you to re-use and tailor arguments for particular situations; claims that are not true in general may be explicitly set in the context in which they are true. You can make your Assumptions explicit in the Goal Structure and state your Justifications at the point you are justifying. You can use contextual information to interweave arguments, making use of claims already substantiated by others. Context is key.

16.3.2 Contextual Relationship

Contextual information is always associated with argument elements using an open-headed arrow, as shown in Fig. 16.7. Contextual reference arrows come from the sides of a Goal or Strategy and point into the side of the Context (or other contextual element). One Context can be at the head of several arrows, i.e. providing contextual information for more than one Goal or Strategy, but check: does it make the diagram clearer, or would a repeated Context be better?

Fig. 16.7 Contextual relationship

Flow of context is essentially sideways from one or more argument elements to each contextual element; and it stops there. There is no context on context. However, contextual information associated with an argument element also applies to the elements into which it is decomposed; context it is inherited.

16.3.3 Context

I introduced Context in Chap. 3; it provides, or references out to, definitions and other supporting material. Represent it as a text box with rounded ends, see Fig. 16.8. If there is a lot of text, use a box with rounded corners (Fig. 16.9).

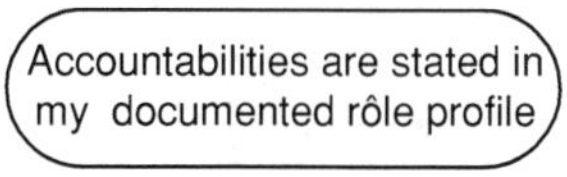

Fig. 16.8 Context

Accountabilities are stated in
my documented rôle profile

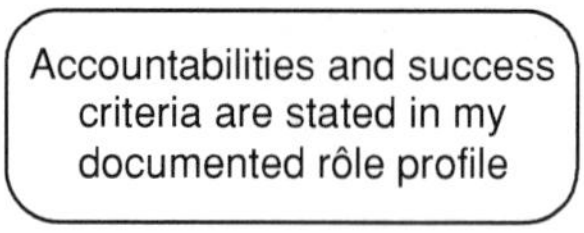

Fig. 16.9 Context for when there is more text to encompass

Accountabilities and success
criteria are stated in my
documented rôle profile

Context applies not only to the Goal to which it is associated (possibly via a Strategy), but also to the argument below it. Therefore, you do not need to restate the Context on Sub-Goals, unless it is required for clarity or for emphasis. For example, for clarity in Fig. 12.5 of Chap. 12, I equip a Sub-Goal with the Top Goal Context from an earlier chapter.

Context lower down the argument structure may also refine that higher up. The whole argument may be about a system; we can use Context to highlight that part of the argument is restricted to a particular sub-system, for example.

Context may also widen the scope of parts of the argument, for example, a safety argument about a device in a particular application may contain sections that are about properties of that device in any (reasonable) application.

16.3.4 *Justification*

Justification is a form of contextual information that I introduced in Chap. 7. It provides extra explanation or rationale; a reason for what you have done. Represent Justification using an ellipse annotated with a J at the bottom right, as shown in Fig. 16.10.

Fig. 16.10 Justification

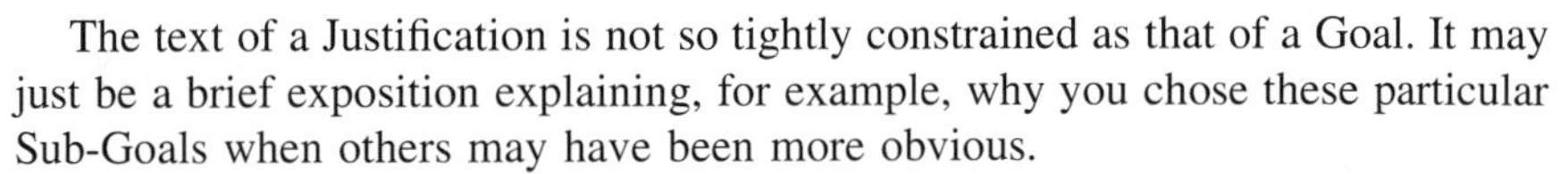

The text of a Justification is not so tightly constrained as that of a Goal. It may just be a brief exposition explaining, for example, why you chose these particular Sub-Goals when others may have been more obvious.

Alternatively, a Justification may contain a simple, "If A then B" argument. Instead, it may point out by reference to an external argument, "Someone Else's Goal", maybe one that has yet to be presented. You can use also Justification in a similar manner to partition your argument, for example so that you can publish it across separate volumes. I give an example of such a partition in Chap. 7.

Justification is contextual information, and so is always associated with the element being justified using an open-headed Contextual Relationship arrow.

Justifications shall only appear at the head-end of arrows, do not hang other contextual information from them. Ideally, for clarity, a Justification should be associated with only one Goal or Strategy, but you may use it to justify more than one if required (using a similar configuration to the shared Context shown in the Problems to Chap. 5).

16.3.5 *Assumption*

I also introduced the Assumption in Chap. 7. It enables you to explicitly state the assumptions that underlie your argument, that provide context for your claims.

Represent Assumption using an ellipse annotated with an A at the bottom right, as shown in Fig. 16.11. Use an open-headed Contextual Relationship arrow to associate it with the argument element about which you are stating the assumption. The arrow tells the reader to read the Goal, or Strategy, in the context of the Assumption. The symbol is, in effect, saying to the reader, "I assume that…", so there is no need to repeat that in the text. The text in an Assumption should be like that of a Goal, i.e. a succinct statement that can be either true or false; the assumption will be that it is true.

Fig. 16.11 Assumption

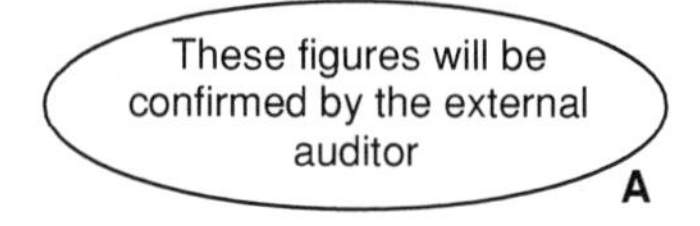

Assumptions shall only appear at the head-end of arrows, do not hang other contextual information from them. You may present an Assumption at the head of several arrows, i.e. providing context for more than one argument element (using a similar configuration to the shared Context shown in the Problems to Chap. 5).

It is important, when finalising your argument, to assess the validity of all the Assumptions; I discuss this in Chap. 7. That Chapter also discusses assumptions used as part of an argument; the example given is argument by contradiction. Do not use the Assumption symbol for this.

That completes the core GSN; the next section looks at the extra symbols need for generic arguments or patterns. But first, here is a diagram showing where all the core symbols may fit in a Goal Structure (Fig. 16.12).

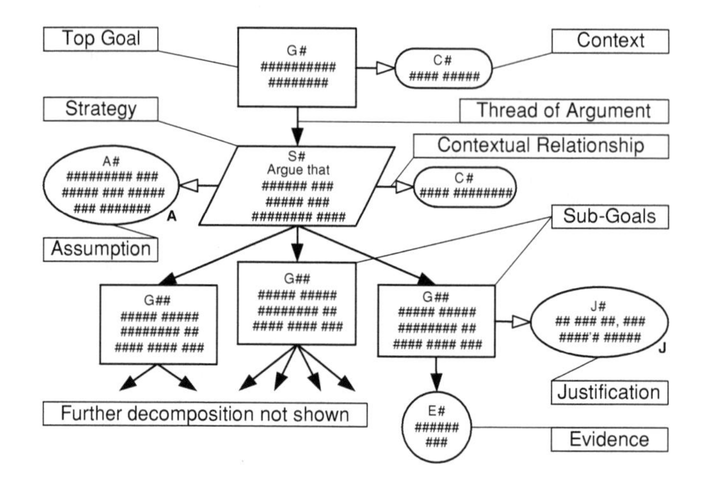

Fig. 16.12 GSN's Anatomy: the naming of parts

16.4 Additional Symbols for Patterns

16.4.1 Preamble

A generic argument, as introduced by Chap. 11, is often called a "pattern", after the design patterns used in some engineering disciplines. It is an argument that applies to a class of things, which you can use as the basis of an argument for a specific example from that class. Producing a particular instance of a generic argument is called instantiation. The text convention is to enclose the name of the class in curly brackets; it can be an arbitrary symbol, such as {X}, but it is better to use more-meaningful terms, such as {Product_Name} or {Vehicle}. Use Context as required to explain what the classes are.

Generic arguments can be developed for many reasons, for example, to encourage consistency across different parts of an organisation, to capture a successful argument for controlled reuse, or to allow third parties to contribute material to your argument. The provision of a pattern does not remove the responsibility to think about the argument that you are making. The generic argument standardises presentation and facilitates review; it does not make your case for you.

16.4.2 Generic Argument Elements

GSN has an extension to the Goal symbol (Fig. 16.13) to represent a Generic Goal, which you may sometimes see referred to as a Goal To Be Instantiated or an Uninstantiated Goal. The Generic Goal is used in just the same way as the normal Goal; it has the same connectivity and text conventions, with the addition of {Class_Identifier}. Chapters 11 and 14 contains several examples. I claim my Generic Goal is true for any member of the set {Class_Identifier}. When I instantiate it in an argument for a particular instance, I will replace {Class_Identifier} with the appropriate element name, and remove the little triangle from the symbol.

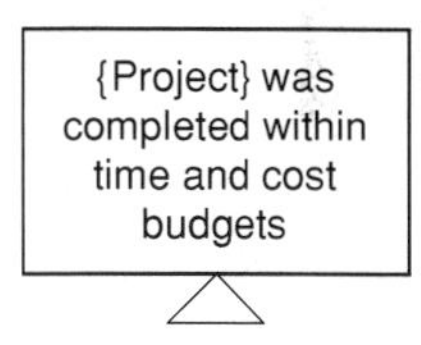

Fig. 16.13 Generic Goal

{Project} was completed within time and cost budgets

Similarly, Generic Strategy (Fig. 16.14) and Generic Evidence (Fig. 16.15) symbols may be used in the same way as you use normal Strategy and Evidence, respectively. They have the same connectivity and text conventions, with the addition of {Class_Identifier}.

Fig. 16.14 Generic Strategy

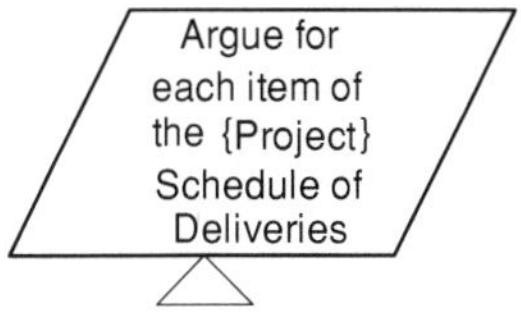

Argue for each item of the {Project} Schedule of Deliveries

Fig. 16.15 Generic Evidence

16.4.3 Optional Relationships

Sometimes a Goal in an argument requires decomposition into a number of instantiations of a Generic Goal. Use a Thread of Argument arrow modified with a 'blob' to show when a number of instances is required, see Fig. 16.16. This symbol is read as, "Supported by zero or more instances of".

Fig. 16.16 Supported by zero or more instances...

If you can state how many instances are required, put the number next to the blob. This quantity is called the cardinality of the relationship. Instead of an explicit number, you could put an n, say, to indicate that you do not yet know the cardinality, see Fig. 16.17. Alternatively you can use a simple expression; such as > 3, meaning more than three.

Fig. 16.17 Supported by n instances...

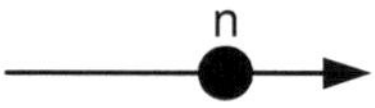

Sometimes a Sub-Goal is only required in some instantiations of the argument, not all. For example you may have special claims about software, but not all the members of the class about which you are arguing have software. Represent this Optional Relationship as shown in Fig. 16.18; do not think of it as a white blob on a black one, think O for Option.

Fig. 16.18 Supported by zero or one instances...

If you are offering a choice between optional Sub-Goals, each relationship has cardinality zero or one, but you have the additional criterion that they are not all zero. Logically, this is an OR relationship; the symbol is shown in Fig. 16.19.

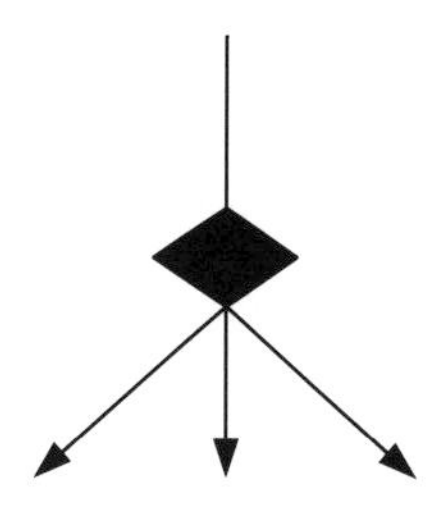

Fig. 16.19 Supported by one or more of these…

In a more complex situation, it many be necessary to choose (at least) three of the options, say. This situation is indicated with a number or simple expression in a similar manner to the Multiple Relationship. This is often called an M-out-of-N, or MooN, Relationship, see Fig. 16.20. It is convention to state both M and N, even if N is obvious from the number of outgoing arrows. Often, the value of N does not actually matter; it is presentation of arguments for the M that has to be achieved.

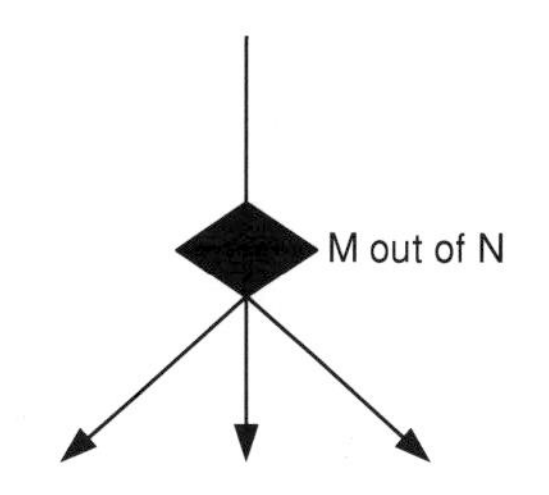

Fig. 16.20 Supported by M of these…

16.4.4 Generic Contextual Elements

You will also need generic contextual elements, if only to define what {Class_Identifier} represents.

Use Generic Context (Context To Be Instantiated, or Uninstantiated Context) in just the same way as you would a Context; it has the same connectivity and text conventions, with the addition of {Class_Identifier}, see Fig. 16.21.

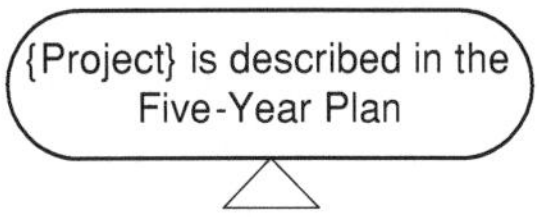

Fig. 16.21 Generic Context

Similarly, there are symbols for Generic Justification and Generic Assumption, see Figs. 16.22 and 16.23, respectively.

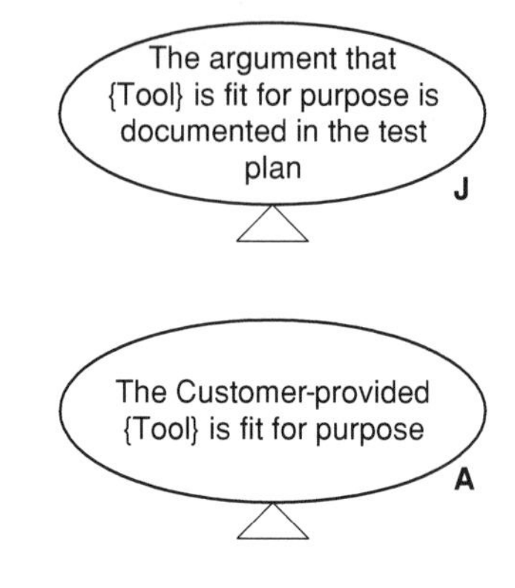

Fig. 16.22 Generic Justification

Fig. 16.23 Generic Assumption

16.4.5 Generic Elements to be Developed

Fig. 16.5 shows the symbol for an undeveloped Goal; that for a Generic Goal that is yet to be developed is an obvious extension. It is shown in Fig. 16.24.

Equivalent symbols for other generic elements that are yet to be developed are sufficiently obvious that they are not illustrated here.

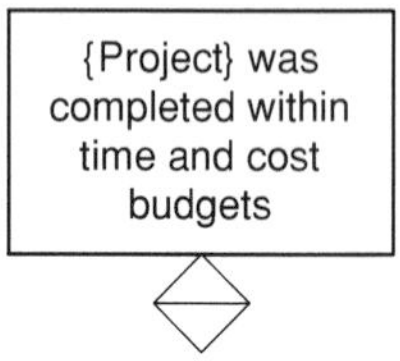

{Project} was completed within time and cost budgets

Fig. 16.24 The Generic Goal to be Developed symbol

16.5 Extended Examples Elsewhere in This Book

- See Chap. 5 for discussion of using an argument to construct checklists that you know to be sufficiently complete and fit for the purpose for which you designed them. This chapter also discusses symbol labelling.
- See Chap. 9 for a discussion of getting your argument reviewed (and making it ready for such a review).
- See Chap. 10 for an example that starts with a question, a conjecture, and then seeks to answer it: is there life on planets in solar systems other than our own? Note that this example overflows into Chaps. 11 and 12.
- See Chap. 14 for an explanation of process argument, a type of argument that is used to show that some process or tool is fit for purpose, or that the evidence produced by a particular use of a process or tool is 'correct'. This material also

includes arguments of competence and some of the difficulties you may encounter in constructing them.
- See Chap. 15 for a brief overview of the extension to GSN that have been developed to support modular arguments - Modular GSN. Even if you do not want to build modular arguments to go with modular systems, you may find this additional notation useful as a means of building a library of argument segments for reuse.

16.6 Question

- Now you are armed with this simple but powerful notation for developing, checking and presenting compelling arguments, who will you persuade first and about what?

16.7 References

I have not explicitly referenced any documents from this chapter, but you may find the following of interest:

Dr Tim Kelly's thesis, in which he develops the concept of using GSN for safety assurance argumentation. Kelly T P: Arguing Safety - A Systematic Approach to Managing Safety Cases (doctoral dissertation), University of York Department of Computer Science, http://www-users.cs.york.ac.uk/~tpk/ listed under Publications 1999

The "GSN Standard", 2010. This document is a *de facto* standard, i.e. it is not recognised by any standardisation organisation, national, regional or international, but it is being developed, by a wide range of contributors in industry and academia, to define GSN and to provide guidance on its usage. http://www-users.cs.york.ac.uk/~katrina/GSN_site/20100517_GSNStandard_v1.0.pdf Note that, despite the filename suggesting that it is Version 1.0, this is an incomplete draft.

Answers to Problems

Abstract This section presents answers to the problems set at the end of each chapter (except for Chaps. 1, 15 and 16; no problems were set therein). Note that in most cases there is no single right answer; an answer is presented below with discussion pointing to other ways of doing things. No external references are made from this section; if a reference is needed it is made from the chapter in which the problem was set.

Answers to the Problems of Chapter 2

The problem asks which, if any, of the given statements could be used as a claim in a Top Goal, i.e. as the basis of an argument. It also asked if any of the others can be reworded such that they can be claimed.

1. "Our Quality Management System is ISO 9001:2008 compliant" is a claim that could be used un-modified as a Top Goal. ISO 9001:2008 is an International Standard for Quality Management Systems; many companies make this claim, but I have yet to see one use GSN to support it.
2. "My Business Plan is complete and ready for review by the Board": At first sight, this looks to be two claims, but that is quite valid. There is nothing to stop your Top Goal claiming that both A and B are True. In this particular case, however, the Business Plan may well be ready for review by the Board, but it will not be complete until that review is finished, you have put in the changes for which they inevitably asked, and you have formally issued the document for use. "My Business Plan is ready for review by the Board" would be better. I expand on this claim in Chap. 5.
3. "This burial site is probably that of Rædwald, King of the East Angles": A valid claim, but it makes you look uncertain of your case. This statement may

J. Spriggs, *GSN—The Goal Structuring Notation*,
DOI: 10.1007/978-1-4471-2312-5,

well be the final outcome of the debate; however, it would be better to start your debate with a positive assertion, "This burial site is that of Rædwald, King of the East Angles".

4. "This painting should be attributed to Albrecht Dürer": again, a valid claim. You could be more assertive, "Albrecht Dürer painted this", but the form of the claim also depends on the purpose of the argument. If you are cataloguing a collection, and have an opinion about the Unknown Artist label given to one of the pictures, use the "should be attributed to" form but, if you found the canvas in your attic and you are trying to sell it, go for the assertive form.
5. "This equipment fulfils the essential requirements of the RTTE Directive" is a valid claim that would normally only be made in a situation where it is known what the RTTE Directive is but, even so, it is better to expand the abbreviation on first use. Alternatively you could identify the directive by number, but it is still not crystal clear (Chap. 3 contains a means of making it clearer). Alternative expressions of the claim are, "This equipment fulfils the essential requirements of the Radio & Telecommunications Terminal Equipment Directive" or "This equipment fulfils the essential requirements of European Directive 1999/5/EC, known as the RTTE Directive".
6. "Beryllia is a carcinogen": Yes, good one; succinct. This is another example where you may need to add information to explain what a carcinogen is—or Beryllia for that matter. What can you assume about what your audience knows? It is difficult for experts from different domains to have a meaningful exchange of ideas because they tend to use different vocabulary, or use the same word to mean different things. You need to be aware of the vocabulary of your intended audience. Are you presenting your argument to inorganic chemists, or to the general public? You will need to expand or contract your descriptions accordingly, but do not put extra information in the Goal; it goes in a new symbol, see Chap. 3.
7. "Hazard Identification and Risk Assessment": It is not unusual to find this one in a system safety argument, but it is not a claim. It is not valid as a Goal. If you find a Goal like this, ask the author what they meant. It may be helpful to suggest an appropriate re-wording; this can show that you have missed the point completely and prompt the author to strive for clarity. They may have meant to claim that, "Hazard Identification and Risk Assessment was done and documented", or "Hazard Identification and Risk Assessment showed that the risk is tolerable", or a host of other things.
8. "Assurance is provided that safety requirements raised on the software are valid": Well, it is a claim, but a bit "round the houses". Presumably, in this case the author will provide an argument assuring us that the requirements are valid. He, or she, will not be providing an argument showing that someone else has already provided assurance that the requirements are valid. We can thus delete the preamble, "Assurance is provided that". Also, note the ambiguity due to the missing article, the claim is "requirements are valid", not

"the requirements are valid". It could thus be interpreted to mean that only some of them are valid; surely, we want all of them to be valid. A better claim would therefore be, "The safety requirements raised on the software are valid". This is still a bit strange, one would normally speak of software safety requirements; so the claim could be succinctly stated as, "The software safety requirements are valid". Half the number of words, but a lot clearer. Note that a similar claim of software safety requirement validity is the subject of a problem set in Chap. 7, "The software safety requirements correctly state what is necessary and sufficient to achieve tolerable safety, in the system context".

9. "The GSN Symbol for a Goal is a rectangle": Yes indeed, but could you argue about it? This is really a definition, rather than a claim. I consider what to do with definitions in Chap. 3. Or, is it a fact, an axiom, or a self-evident truth? I deal with those in Chap. 7.
10. "The colour of the sky" is not a claim. It could be made a claim by choosing an actual colour, "The sky is blue". Even now, this is not a good start for an argument; sky colour is not an invariant, it is different for night and day, for example. It is different in different places; it depends on the weather, and so on. By the time you have finished preparing your argument the sky may have turned grey. Think carefully: what is it that you want to argue, who is it that you want to persuade, and why? Once you have answers to these questions you will be well on the way to expressing a good Top Goal, "The daytime sky seen from the surface of Mars on a clear day is yellow–brown", for example.

Answers to the Problems of Chapter 3

1. If this were a real example, I would have changed the claim to, "Preventative Maintenance Procedure PMP5 is fit for purpose" and would have provided more detail as to what "keeping the power supply running" actually means. But that was not the question; using the information provided, Fig. A.1 is a potential solution. I could have split the left-hand Context into two; one to specify what PMP5 is and the other to say where to find it. Of course, in some environments, "PMP5" may be sufficient reference, making it unnecessary to state where it is documented. Note the Context specifying the factory power supply; this is building on the definition of purpose, rather than the claim itself, but it is shown linked to the Goal, not the Context. Context does not have context of its own (unless it is an external document referenced in the Context itself). It may have been clearer to combine this pair into one Context. In this solution, I have assumed that the references specify versions, issue states, etc. If they did not, I would add a Context to say, for example, "This argument is for PMP5 Issue 2 as applied to Power Supply build-state 7.2".

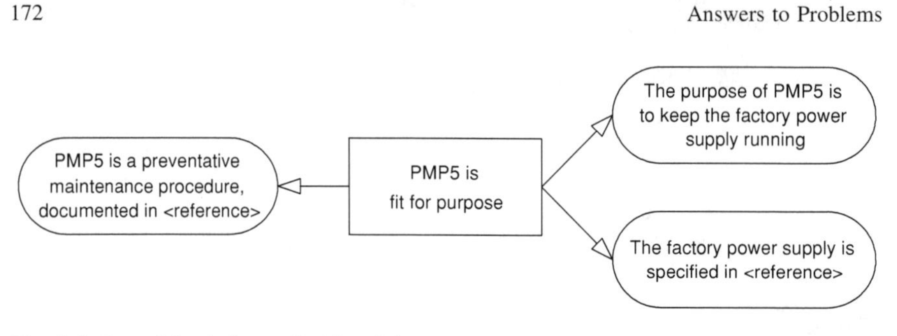

Fig. A.1 Potential solution to Problem 3.1

- Note that this last problem is not as contrived as it may seem. It is not unknown for preventative maintenance procedures to be changed, or dropped altogether, by people who do not know what the purpose of the procedure was. I have also encountered the converse; I was told of a procedure that was being regularly carried out to check a piece of equipment that was only there in case of an incident with some machinery that was no longer in service (in fact no longer there).

2. This solution is very similar in concept to the previous one; in this case, allegedly, the Butler did It, so we need to specify both Butler and It, see Fig. A.2. Note that I have taken the fictional case reference and emphasized it to form a "tag". If you have a set of small arguments, such tagging makes it easier to find the one you want.

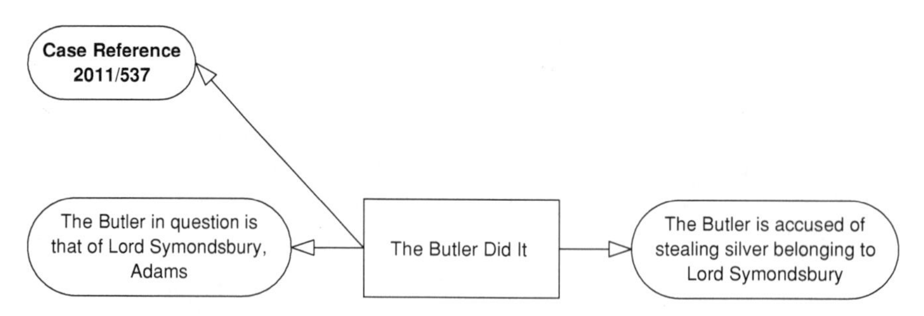

Fig. A.2 Potential solution to Problem 3.2

Answers to the Problems of Chapter 4

1. The expected conclusion is that vinegar is an acid, as shown in Fig. A.3.

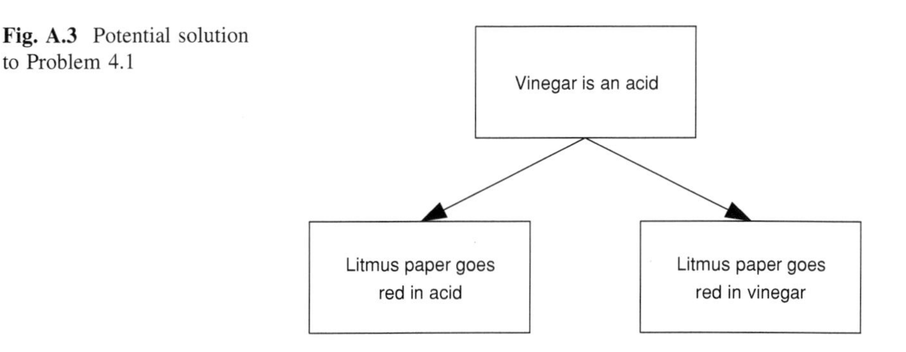

Fig. A.3 Potential solution to Problem 4.1

2. As well as a Context to present the fletchings definition given in the question, I have added one to explain tumbling, see Fig. A.4.

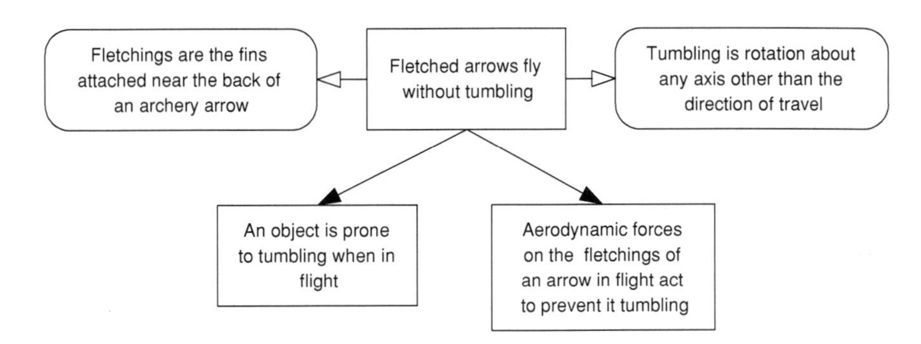

Fig. A.4 Potential solution to Problem 4.2

3. Note that, in Fig. A.5, I have included a defining Context for clarity.

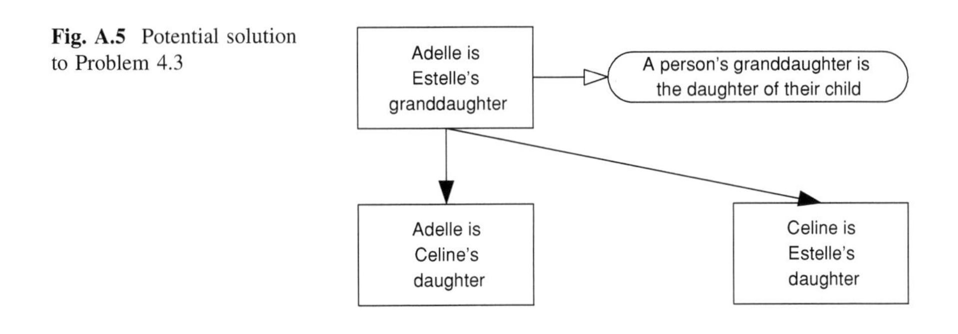

Fig. A.5 Potential solution to Problem 4.3

Answers to the Problems of Chapter 5

1. From the names, this appears to be a French family so, in Fig. A.6, I have phrased my claims in the French manner, rather than using apostrophes.

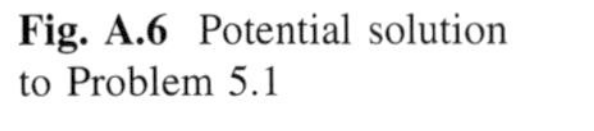
Fig. A.6 Potential solution to Problem 5.1

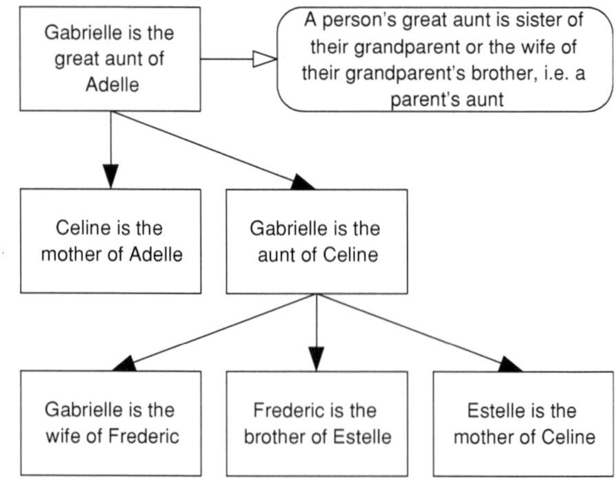

2. I revert back to using apostrophes in Fig. A.7, in line with the phrasing of the Goal re-used from Chap. 4. These two examples show that there is more than one way of constructing an argument; we should strive to examine the alternatives and pick the most compelling or the clearest for use.

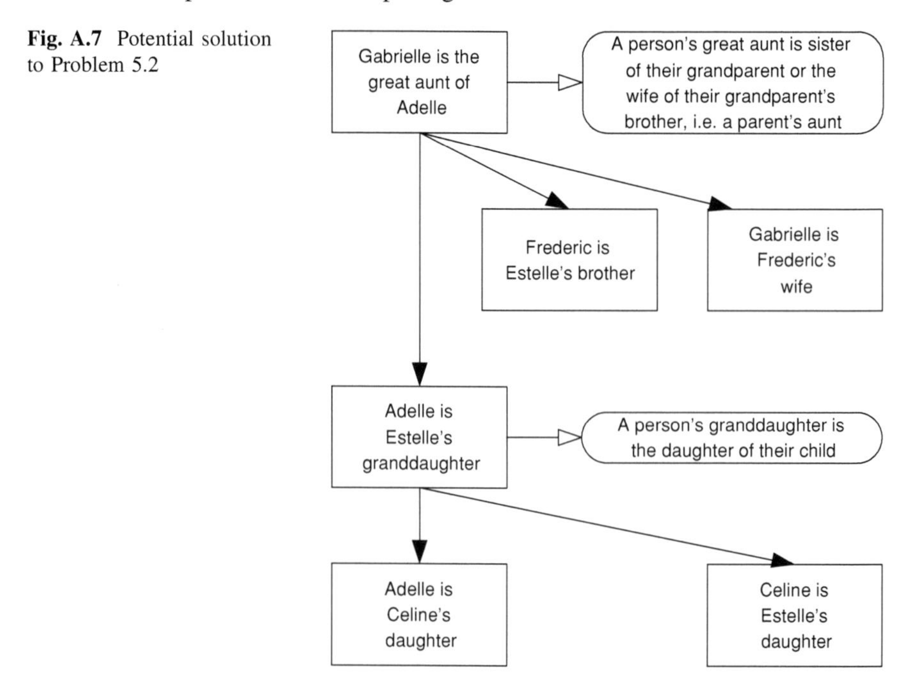

Fig. A.7 Potential solution to Problem 5.2

3. Context applies to the Goal to which it is attached and all of its Sub-Goals. In this example, it applies to all the Sub-Goals of a Goal. It is therefore reasonable to apply it to the Goal itself. As a general rule, apply Context as far down the Goal Structure as practicable; it is then clearer for the reader to understand your meaning. Invoking a standard, for example, at the top of the structure just adds confusion if the subject of that standard does not appear until two levels down in the argument structure. Invoking the standard and introducing the subject at the same level immediately sets the context and aids comprehension.

Answers to the Problems of Chapter 6

1. A Context is missing from the figure. It is needed to state that the three sub-divisions, as represented by the Strategies, cover all the requirements.
2. We can include Context in the Sub-Goal version to show that the three sub-divisions cover all the requirements, as shown in Fig. A.8. Alternatively, we could add a fourth Sub-Goal to argue that there are only these three types of requirement.

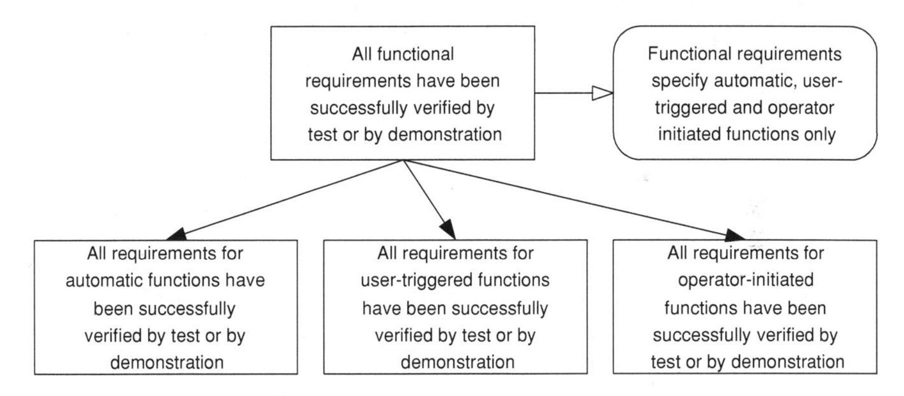

Fig. A.8 Reducing span by using Sub-Goals to add levels

Answers to the Problems of Chapter 7

1. Assumption A6.3a includes the preamble "It is assumed that"; the symbol is to be read as meaning that, so it is unnecessary to state it. It should have been just "Vehicle traffic is in the junction". Examine that statement; whether it turns out to be true or false, it has no impact on the truth of the Goal. A6.3a is thus an unnecessary Assumption and should be removed. A6.3b contains an external reference to supporting information; it should therefore be a Context. I already

have a Context, so I have re-labelled it C6.3a and transformed A6.3b into C6.3b. A6.3c appears reasonable, but as applied to the Goal, rather than the Strategy. I propose re-drawing the argument fragment as in Fig. A.9.

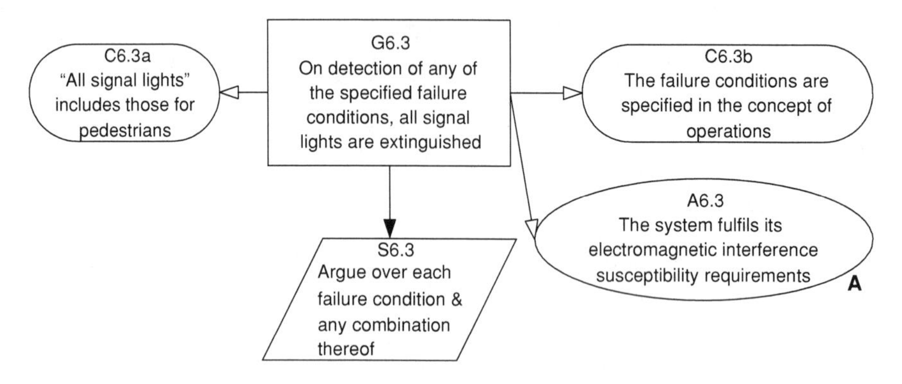

Fig. A.9 One Assumption remains

2. The basic answer is shown in Fig. A.10, in which I added a Context to say where the Goal statement came from. I should also have added Contexts to say what the system is and where the various requirements referred to are specified. For the next step, Fig. A.11, I add those and use the text of the notes (but I do not need to say "It is assumed that…" in the Assumption symbols).

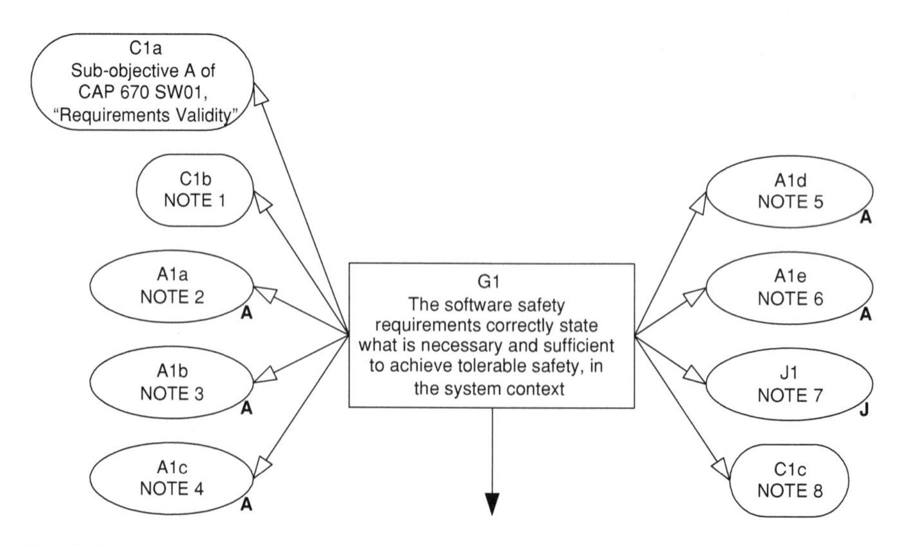

Fig. A.10 Basic answer using the suggested shorthand

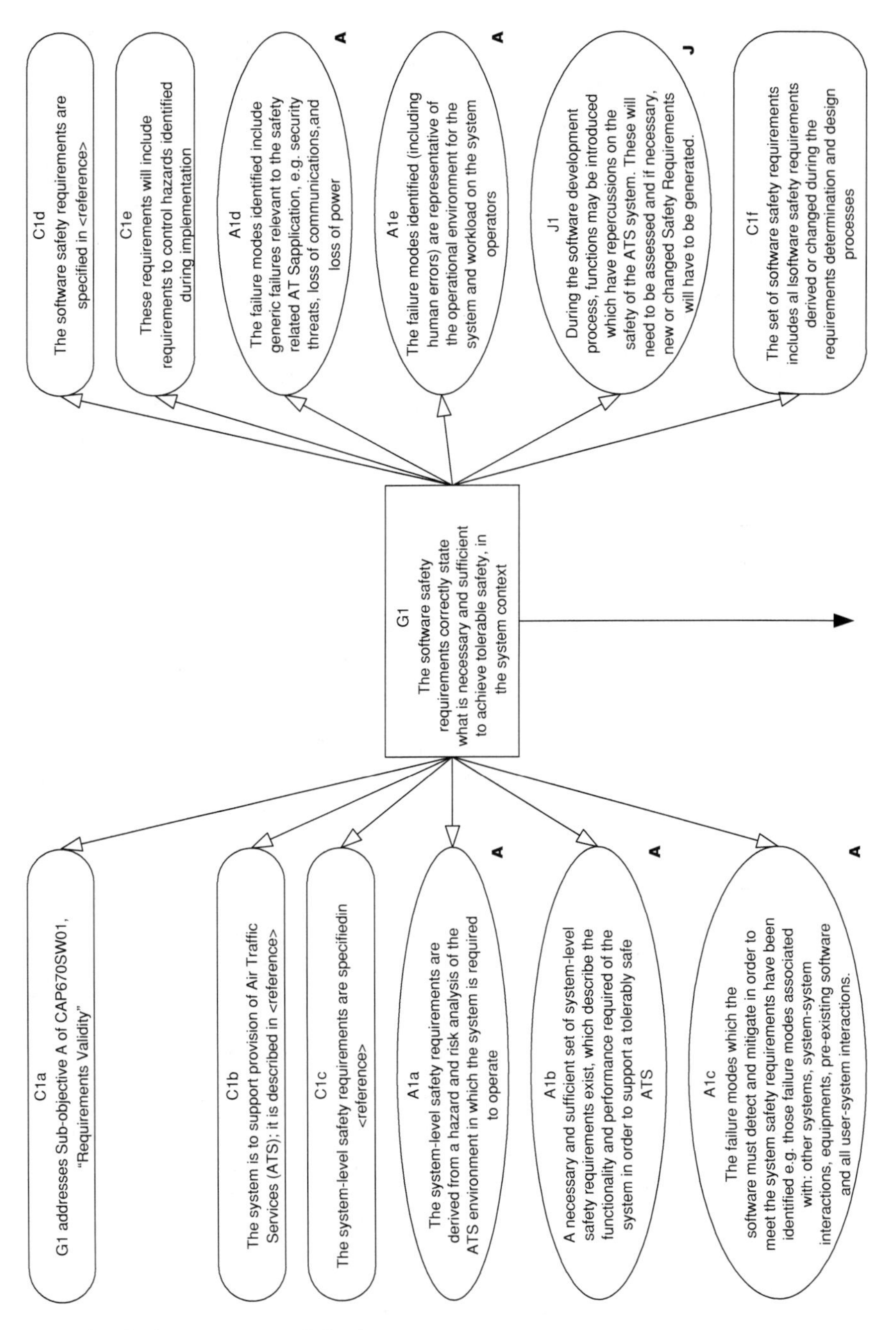

Fig. A.11 Basic answer expanded using note text

3. The problem is that we have been too literal in our capture of the material and have missed a significant point. This is a case in which a Customer (actually the Customer's Regulator) has told us what to argue; they want us to demonstrate that the claim is true for our software-based system. They stated assumptions, but these are not the conditions in which the claim is true, rather they are the conditions for it to be the right claim. We must not capture them as Assumptions; we must show in our argument (Fig. A.12) that they are valid.

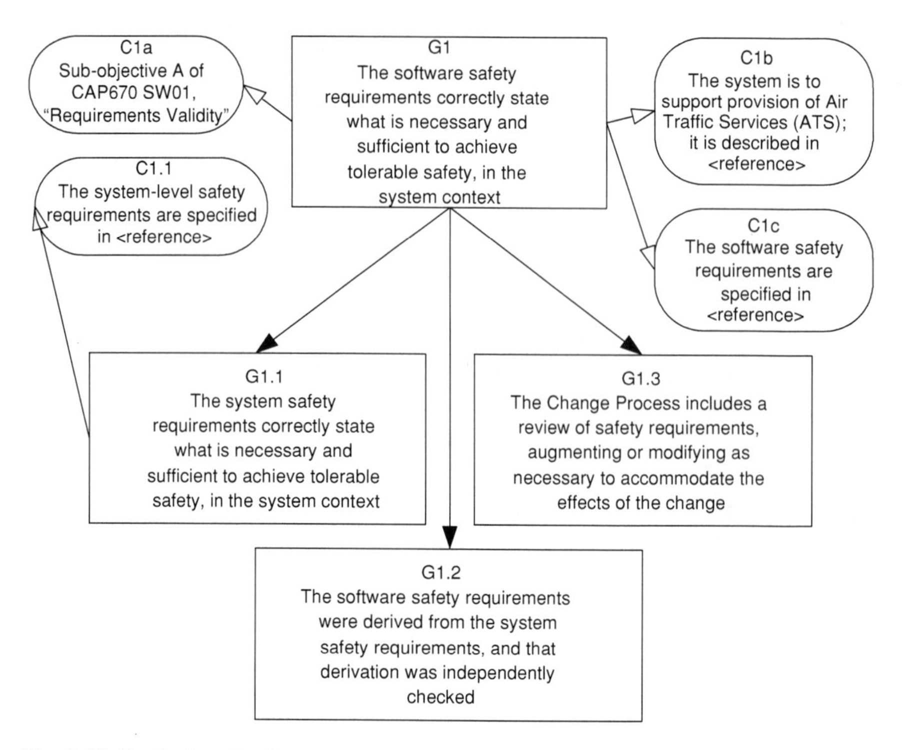

Fig. A.12 Revised top-level argument

It may have been better, rather than to have detail of a low-level process highlighted right at the top of the argument, to have had a Justification in place of Goal G1.3. That Justification would have the same wording as the Goal, but it would also cross-refer to a lower-level Goal that provides the argument.

I have not captured all the Customer's assumptions in the decomposition shown in Fig. A.12; to do that I will need to decompose Goal G1.1 another level, see Fig. A.13.

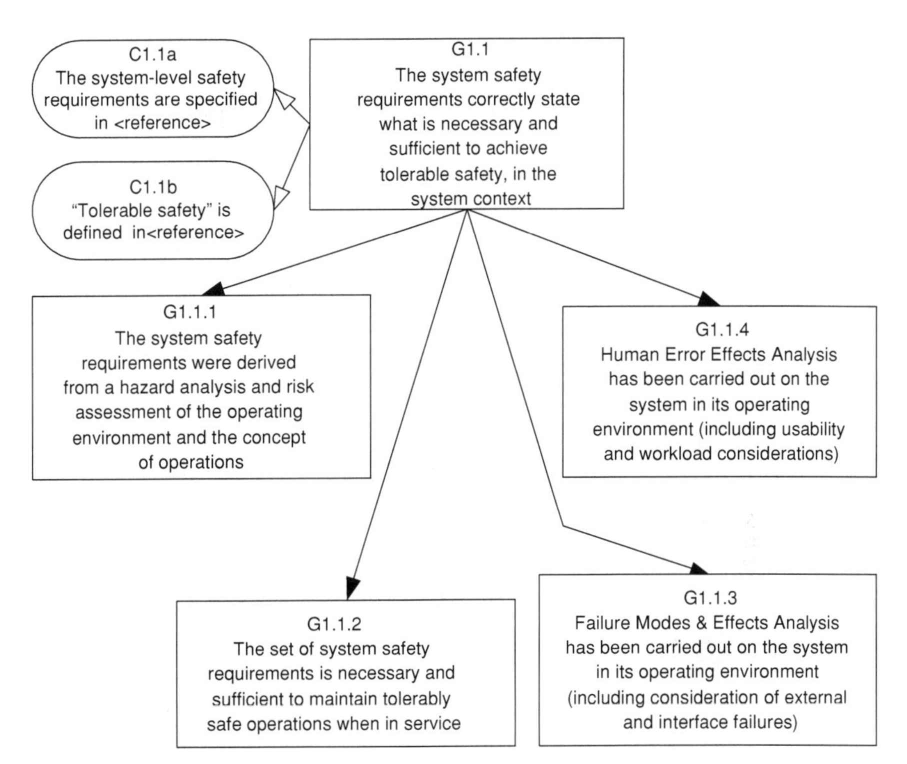

Fig. A.13 Decomposition of system requirements Goal

Answers to the Problems of Chapter 8

1. It is conceivable that, if you were to adopt the S convention for numbering evidence, you could have two entities in your argument with the same number, the other one being a Strategy. This does not really matter to the reader, as it is clear from the geometry and the syntax which is meant; it can, however, cause confusion in review or challenge. For example, someone may have written, "What is the justification for S2.1.2?" Do you have to explain why you used that Strategy, or why that Evidence is pertinent?
2. We can use the same structure as we did for the Methuselah Report, as shown in Fig. A.14. I would hope to see a bit more information in a real argument, see Problem 3.
3. This part of the argument would be more (or possibly less) compelling if the actual result of the analysis were declared. If the report predicts a failure rate of once in a hundred thousand hours, I will be more confident that the claim is true than if it had predicted exactly once in ten thousand hours. Of course, if the

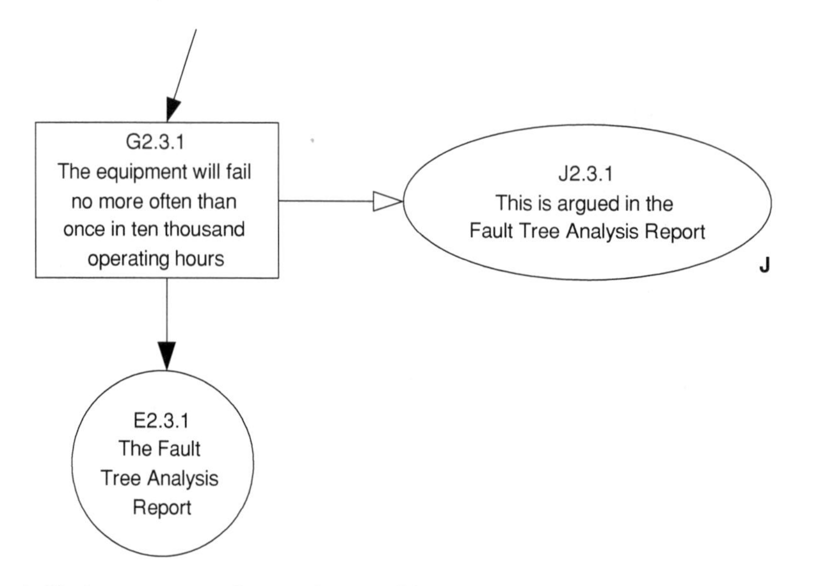
G2.3.1
The equipment will fail no more often than once in ten thousand operating hours
J2.3.1
This is argued in the Fault Tree Analysis Report
J
E2.3.1
The Fault Tree Analysis Report

Fig. A.14 Argument part redrawn using core GSN

report were to predict once in a hundred million hours, I could be less confident, thinking instead that the people who prepared and approved the report may be kidding themselves!

Another point of concern here is whether Fault Tree Analysis is valid in this situation. The logical structure of such an analysis is universally applicable to causal systems but, like in the Reliability Block Diagram example in the main body of the chapter, the failure rate calculations depend on assumptions that may not hold for this particular equipment. The use of the technique should be justified, as should the competence of the analyst, the suitability of any tools used to produce the results, and the provenance of the data.

The argument segment, shown in Fig. A.15, also illustrates another problem with evidence in some contexts: the potential for mismatch of units. When I made the claim, I expressed the target in terms of hours but, when the analysis report arrived, it gave the result in years. Although the widespread use of standard "SI Units" has reduced such problems, there are situations in which other units, such as Knots, have been retained. I suggest either re-expressing the target in new units using a Context, as shown, or using a Justification to declare that the target has been met, in this case because failures are predicted to occur at least eight times more infrequently than required.

Also, is it sufficient to point your readers to a, potentially very large, report as evidence of a parameter attaining a target? It would have been better if I had identified the pertinent section of each report as Evidence.

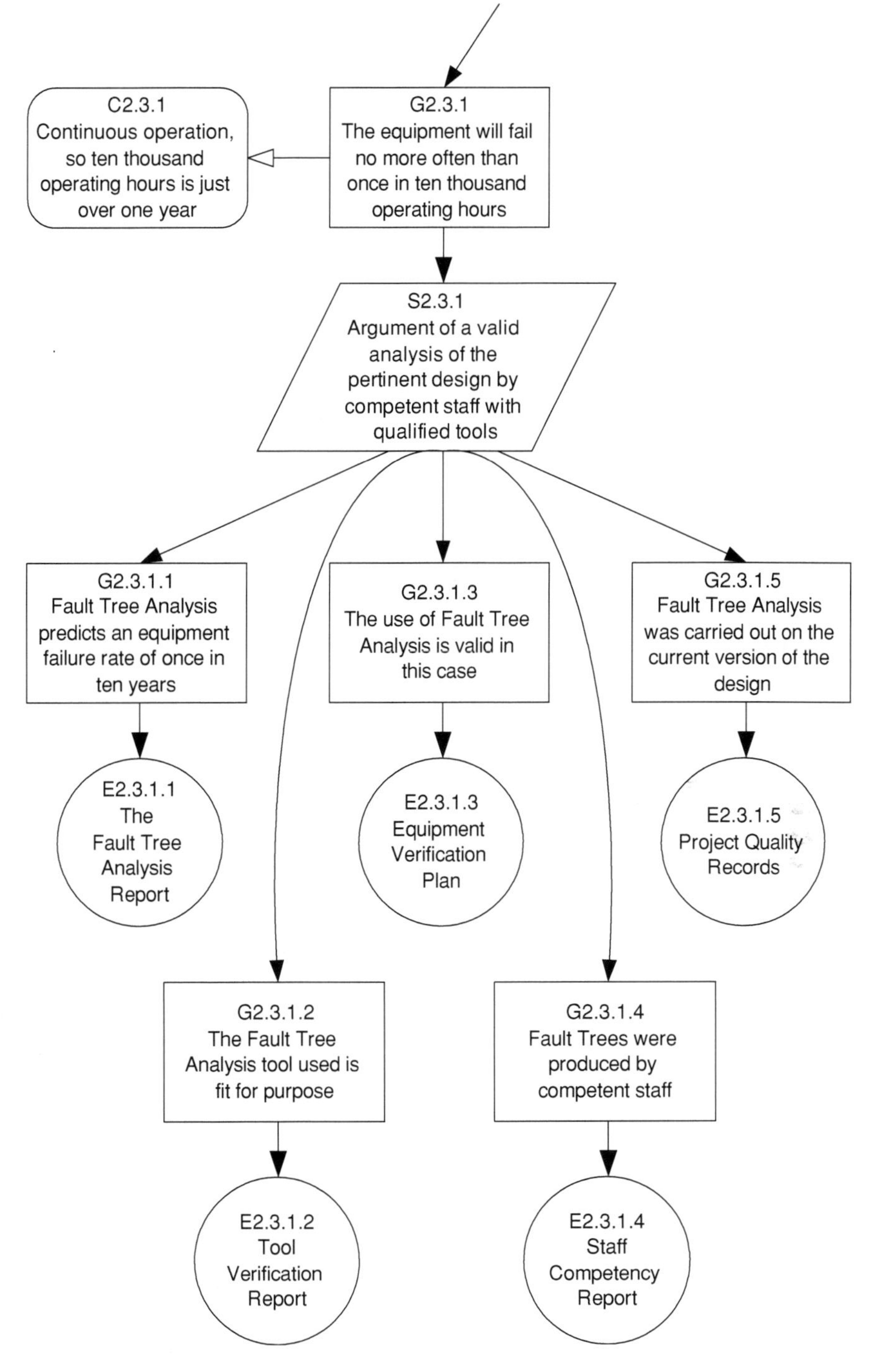

Fig. A.15 Argument augmented with new Sub-Goals

Answers to the Problems of Chapter 9

1. If you were to use the Goal to be Developed symbol to indicate that a Goal is developed later on in the same document, it would cause confusion for the reviewer, who will not be expecting to follow that particular chain of thought further. It would cause even more confusion if your argument were to include genuine undeveloped Goals. If you wish to include navigation information on the diagram, use a Label; otherwise, put it in the text below the diagram. Use a tabular or bulleted format if there are several such Sub-Goals.
2. I am sorry; I cannot give you an answer to this one. What you need to produce is a personal checklist that includes the detail you need. If you are naturally very methodical and/or succinct and accurate in what you write, you can prune those areas of the checklist, but you may need to enhance other areas… You may also wish to add additional questions once you have read Chaps. 12 and 13 on the problems of evidence collection.

Answers to the Problems of Chapter 10

1. Although now a deprecated symbol, this seemed an ideal opportunity to use a Model, see Fig. A.16. In a real argument, I would actually have included it lower down the Goal Structure, in the decomposition of Goal G3.3…

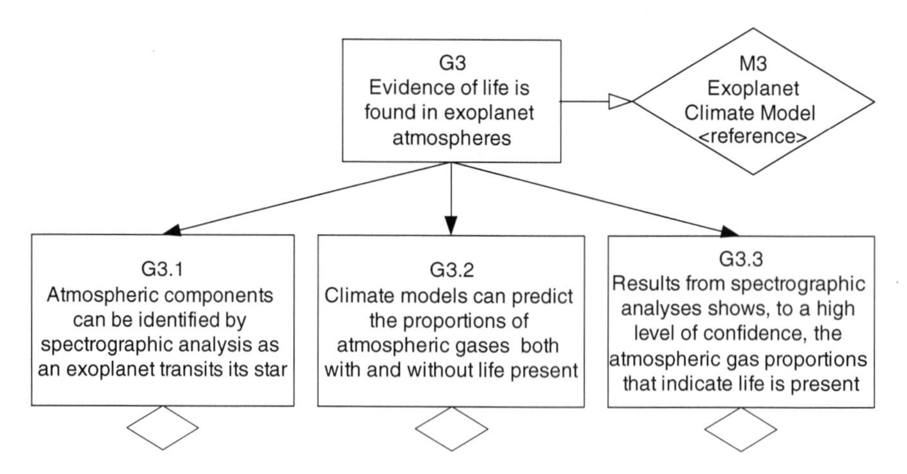

Fig. A.16 Potential solution to Problem 10.1

2. My proposed solution is set out in Fig. A.17.

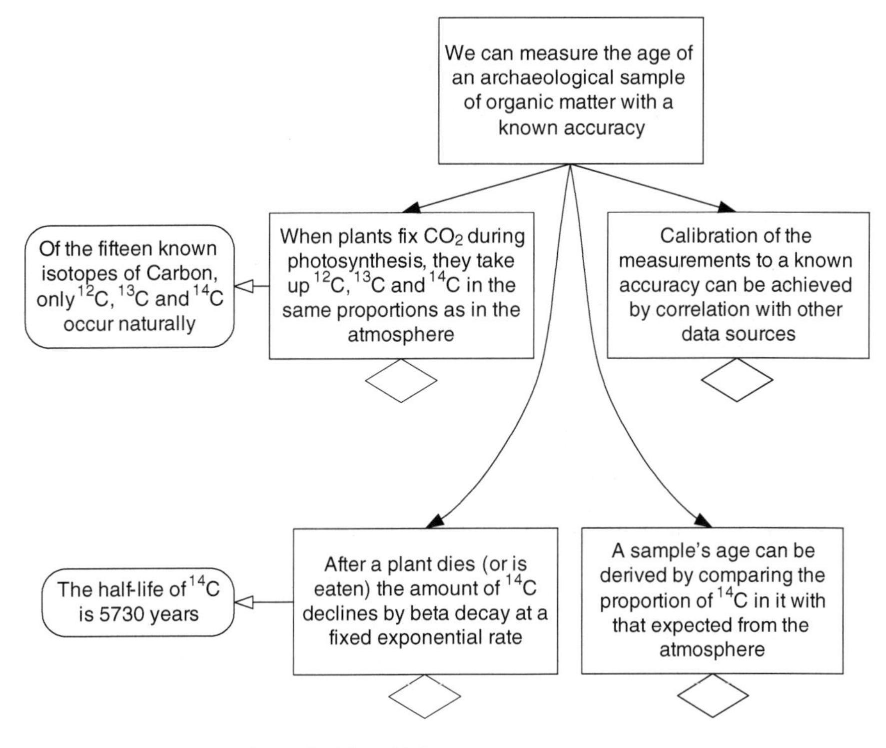

Fig. A.17 Potential solution to Problem 10.2

Answers to the Problems of Chapter 11

1. This is my suggested decomposition; I have split it into two diagrams (Figs. A.18 and A.19) for ease of fitting it on the page. If there are tasks to perform, they must be identified and specified so that training can be provided. That is not enough; we also need to run some evaluation trials so that Users can raise any concerns and we need to monitor performance when the system is in service, and feedback any problems encountered to the System Authority.

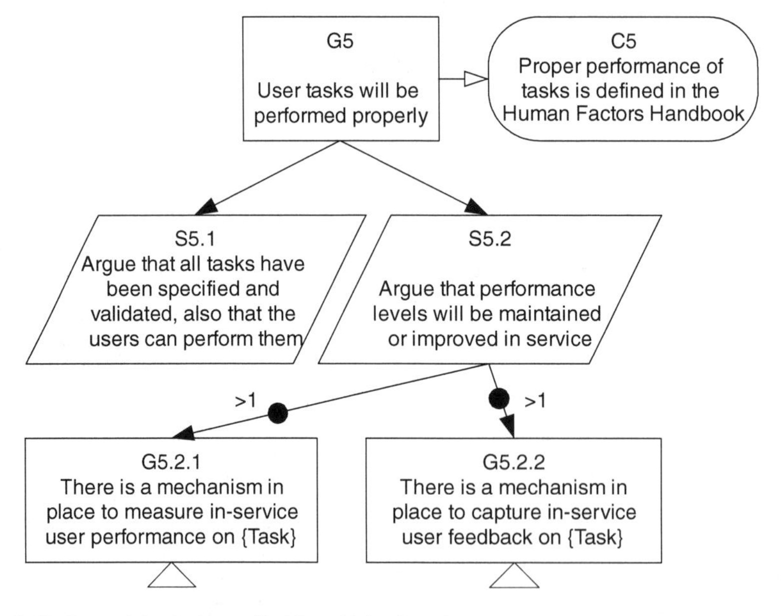

Fig. A.18 Potential solution to Problem 11.1, view 1

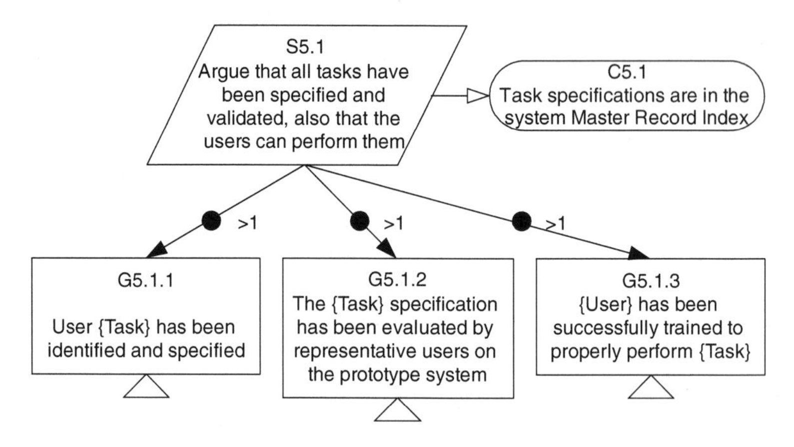

Fig. A.19 Potential solution to Problem 11.1, View 2

Answers to the Problems of Chapter 12

1. In the absence of any prior use data from other systems, I have to depend on the experience gained with the component as it was deployed in the system throughout testing and evaluation. The usage environment should be very similar to that which will be experienced in operation, indeed I have had to argue that elsewhere in the assurance to justify the use of the test set-up. In a

real argument I would have bought out the number of hours of use the component had been subject to. Here, in Fig. A.20, I have left the reader to get that from reading the Evidence reports.

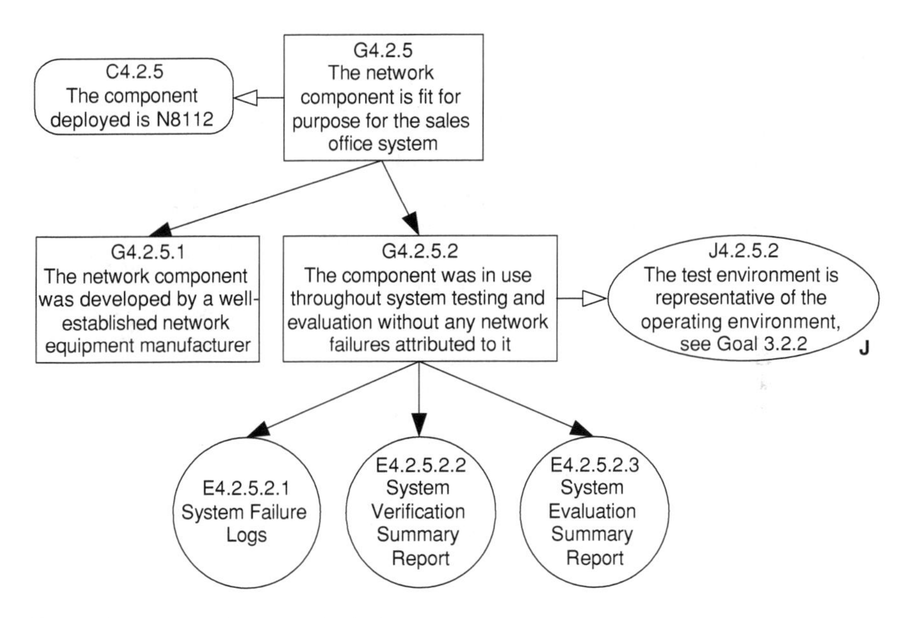

Fig. A.20 Argument based on the deployed component

Answers to the Problems of Chapter 13

1. The key point is that, in the fish example, the counter-evidence challenged the Top Goal directly, whereas the test failure challenged a Sub-Goal way down the Goal Structure. In general, if a small part of your argument is refuted by counter-evidence, it may only be that part that is wrong; it may still be possible to argue for the Top Goal successfully. Look for a work-around, or (and this is the better option) develop a different way of supporting your claims.
2. The example found in the fisherman's catch was dead; therefore the Coelacanth is extinct, but the extinction event was a bit more recent than the previous estimate of sixty-five million years ago. OK, a silly example, but if you are a reviewer of arguments, this is the sort of thing you should be looking out for.

Answers to the Problems of Chapter 14

1. I have spread my suggested solution across three figures; one for the Top Goal decomposition (Fig. A.21) and one each for each of the Strategies (Figs. A.22 and A.23).

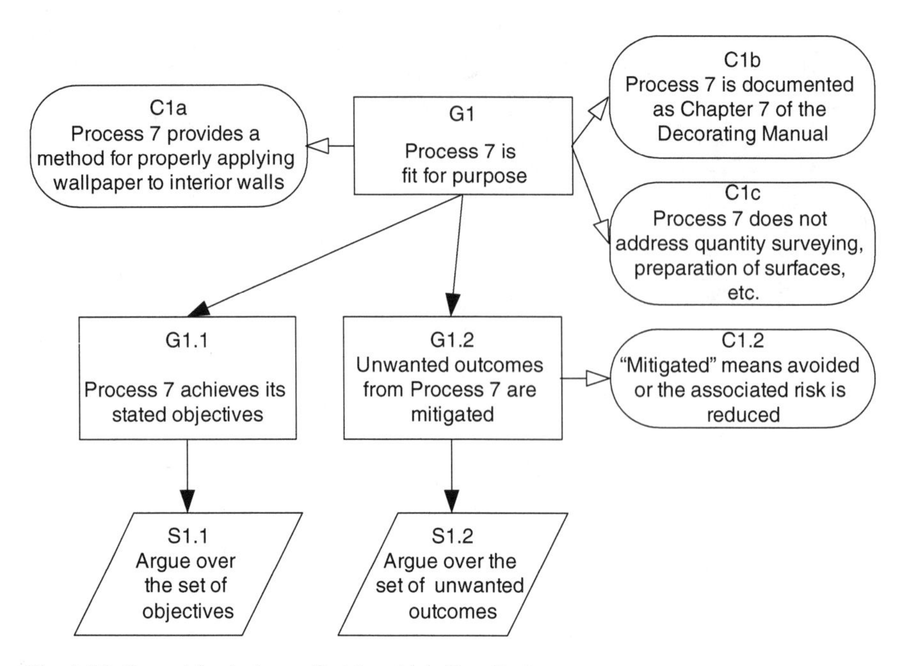

Fig. A.21 Potential solution to Problem 14.1, Top Goal

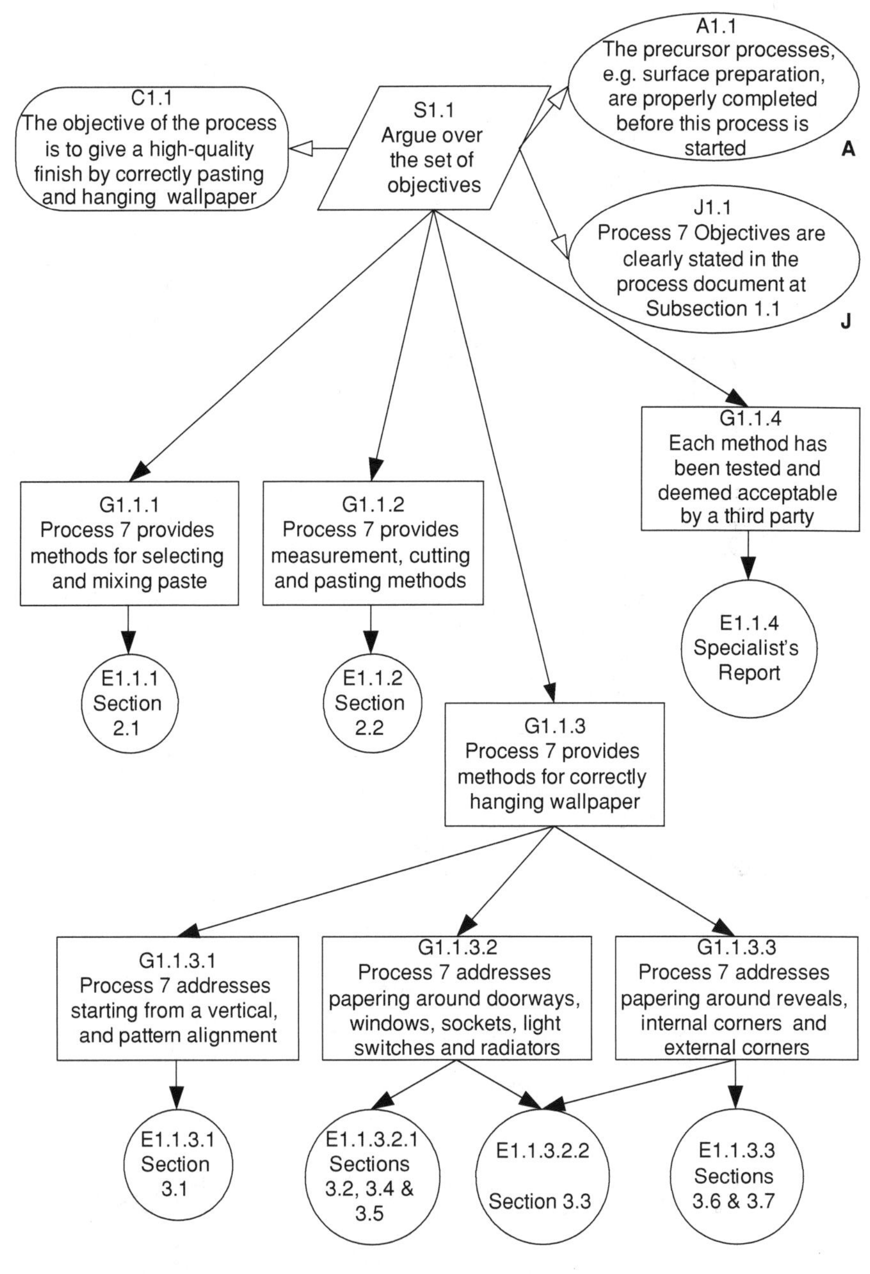

Fig. A.22 Potential solution to Problem 14.1, S1.1

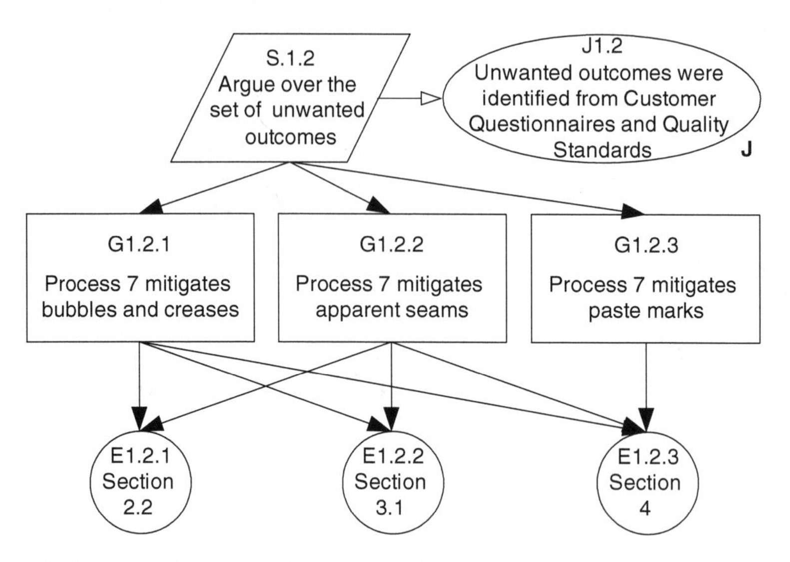

Fig. A.23 Potential solution to Problem 14.1, S1.2

2. I asked you to modify the pattern in Fig. 14.1 to argue for the outputs of a computer-based tool; see Fig. A.24 for my solution. In practice, such a tool would be used to automate part of an existing process; if so, you should augment the process argument with the tool argument, rather than replace it.

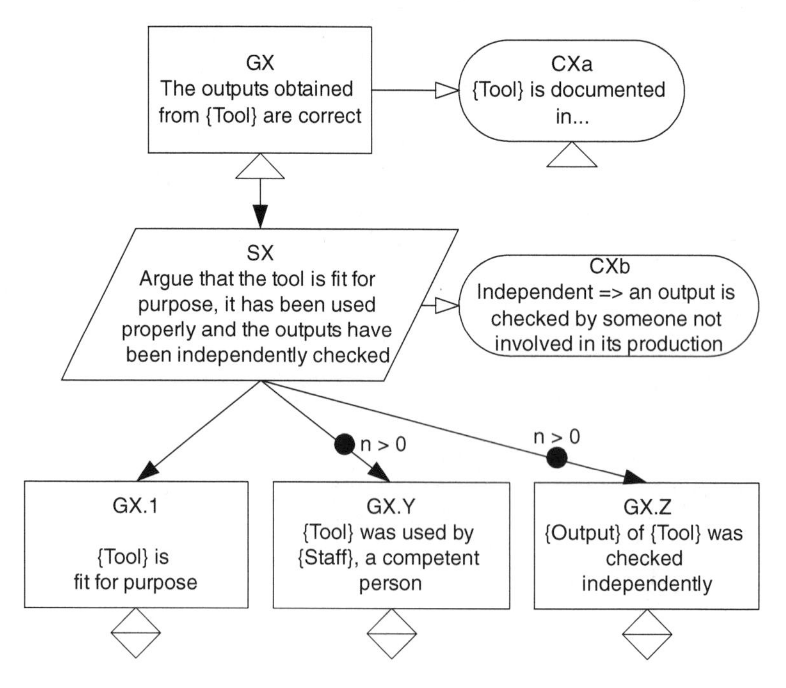

Fig. A.24 Potential solution to Problem 14.2

Index

J. Spriggs, *GSN—The Goal Structuring Notation*,
DOI: 10.1007/978-1-4471-2312-5, © Springer-Verlag London Limited 2012

PGSTL